Getting to Know Your Cells

Leslie Saucedo

Getting to Know Your Cells

Illustrations by Maria Jost

 Springer

Leslie Saucedo
Biology Department
University of Puget Sound
Tacoma, WA, USA

ISBN 978-3-031-30145-2   ISBN 978-3-031-30146-9   (eBook)
https://doi.org/10.1007/978-3-031-30146-9

Cover illustration: © *Anusorn Nakdee, Shutterstock 2023*, All Rights Reserved
Illustrations by: Maria Jost, Tacoma, Washington

This Springer imprint is published by the registered company Springer Nature Switzerland AG
The registered company address is: Gewerbestrasse 11, 6330 Cham, Switzerland

Acknowledgments

Many thanks to all those who mentored and inspired me to become a teacher-scholar, the friends and family who encouraged and supported me along the way, and the students who continuously motivate me with their curiosity and enthusiasm to engage.

Contents

1   Introduction   1

2   Red Blood Cells   7

3   Stem Cells   13

4   Epithelial Cells   19

5   Hepatocytes   25

6   Fibroblast Cells   31

7   Neurons   37

8   Muscle   43

9   Endocrine Cells   49

10   White Blood Cells   55

11   Oocytes   61

12   Sperm   67

13   The Immune System   73

14   Cancer   79

15   Conclusion   85

Sources   87

Index   95

1

Introduction

In theatrical productions, Frankenstein proclaims "It's alive!" upon the animation of his monster – a life literally stitched together from pieces of dead humans. While rousing the dead is eerie enough, there is something additionally unnatural about a new life formed from a novel assemblage of old parts. Yet all life is exactly that. Perhaps most beautifully captured by Carl Sagan when he said "We're made of star stuff," it is also true that we are made of worm feces. Similar to the body parts of the monster, neither star stuff nor worm feces are alive. In addition, many of the parts that we are currently using as we live today are physically not the same ones we were using last week or will be using next year. While you can look at your hand and consider that you have harbored the same unique fingerprints since your birth, the parts that form those fingerprints have been exchanged many times, though in a far more subtle way than cutting and sewing limbs.

The assembled fragments of Frankenstein's monster are visible to the naked eye, but star stuff is not. In this book, we will focus our attention on a place between those two levels, the cell – the smallest part that is actually alive. The term "cell" was first used by Robert Hooke as he described cork using an early microscope in 1665. Interestingly, Hooke was actually visualizing the dead, block-shaped remnants that surround and serve as external protection to living plant cells – these walls are made of cellulose and are a large source of fiber in human diets. The empty space within the blocks was where cells used to live. Living cells were documented by Antonie van Leeuwenhoek a decade later, which revealed cells that were not simply static blocks. Hooke's microscope magnified objects up to 50-fold, and van Leeuwenhoek's microscopes approached a magnification of 300-fold. Today's microscopes can provide a magnification of about 1000-fold on living samples and over 1,000,000-fold on samples that are dead. Samples used to study cells from humans are typically obtained from biopsies. The cells can be observed in the context of a tissue sample taken (after very thin slices are prepared), or the cells can be grown outside of the donor using a method called tissue culture. Tissue culture involves growing cells in plastic dishes containing nutrients and the appropriate environmental conditions. It is a very powerful way to manipulate one variable at a time and then witness how cells respond. However, cells in culture often exhibit properties not normally observed in a living organism or in vivo, likely because of the loss of tissue architecture and the failure of the culture environment to accurately reproduce the more complex physiology of an intact organism.

As microscopy improved, it became apparent that cells were very complex yet organized. What cells "do" is performed by molecules, such as proteins, that are too small to see directly in living cells. Importantly, a method called fluorescence microscopy allows these molecules to be seen indirectly. In short, the molecule of interest is attached to another molecule that glows under certain wavelengths of light (Fig. 1.1). In this way, the location and movement of individual molecules can be tracked even in living cells. A good analogy is that you cannot directly see a person walking on the sidewalk from an airplane 10,000 feet in the air, but you would be able to see a powerful light that the person is holding in the dark. This type of microscopy dramatically

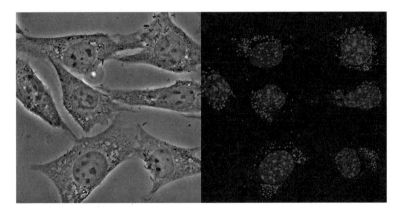

Fig. 1.1 *Detecting cellular molecules.* Cells are mostly translucent, and microscopy using visible light (left) can only reveal differences in density – often only the nucleus is distinguishable. Fluorescence microscopy (right) reveals the location of molecules by attaching pigments that glow under UV light. In this case, DNA is being detected in the nucleus (blue) and lipid droplets in the cytoplasm (red). (From https://analyticalscience.wiley.com/do/10.1002/imaging.5611)

revealed the very dynamic processes that take place within cells, including molecules motoring along subcellular highways (here is an example: https://www.youtube.com/watch?v=PqS1ePevIrU).

Many living organisms consist of just one cell, making them hard to notice and their abilities difficult to appreciate. Some examples are bacteria, algae, and yeast. When we do detect them, it is because they have replicated themselves many times such that hundreds to millions of cells have congregated in a discrete location. Single-celled organisms are sometimes described as simple, but all living things must accomplish very complicated feats, such as harvesting energy, responding to their environment, and reproducing. Single-celled organisms have existed on Earth for about two billion years before multicellular organisms evolved and continue to inhabit environmentally extreme locations on Earth that multicellular organisms do not. The evolution of multicellular organisms created the opportunity for some cells to specialize – that is, to become experts at a subset of life functions. This book will primarily examine how specialization has played out for several of the cell types found in mammals, including us. But the general idea of the benefits of specialization is nicely illustrated by a protist called *Dictyostelium*, also known by the common name slime mold. *Dictyostelium* can live its life as a single-celled organism that crawls around in the dirt, eating bacteria. However, if food sources become limited, single cells begin to aggregate together to form a slug-like organism. The slug crawls toward light, and the cells rearrange to grow a stalk upward, culminating in a tip structure that bursts to disperse some of the cells to new areas. This process enables some cells to live while sacrificing many others (Fig. 1.2, https://www.youtube.com/watch?v=bkVhLJLG7ug).

Cells working together is paramount to the success of multicellular organisms – but unlike *Dictyostelium*, our cells become so specialized that they do not have the option to set off on their own. Strikingly, cells in multicellular organisms have even evolved the ability to actively self-destruct as part of organismal development and maintenance of adult form. A malfunction of this genetically programmed cell death can be seen in humans who retain the webbing between their fingers during embryo development. Several other measures are in place to compel individual cells to cooperate and sacrifice for the benefit of the whole organism. Disabling these barriers can allow a cell to be unconstrained and become cancerous.

You may know that you started out as just one cell – called a *zygote,* which formed upon the union of a sperm and an egg. That zygote contained the deoxyribonucleic acid (DNA), which has served as a set of instructions to build approximately 37 trillion cells that you now call you. Intriguingly, DNA is (mostly) faithfully copied and passed on to each new cell. Yet about 200

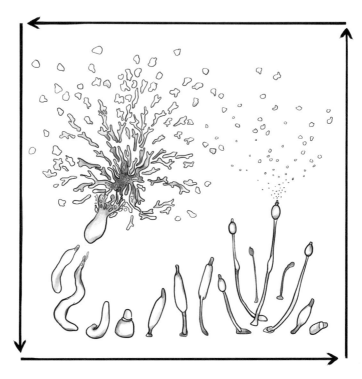

Fig. 1.2 *Dictyostelium: an organism that can vacillate between a single-celled and multicellular life.* Under nutrient-limited conditions, independently living cells will congregate by the thousands to form a multicellular slug. Upon crawling to a new location, the slug transforms to form a stalk and bulb structure, which can then release individual cells into the wind and, hopefully, to a location with better resources

different types of cells are produced using these same instructions. This is because each cell type varies on the subset of instructions that it "reads." Some instructions are universal and will be used by all cells to accomplish the base functions needed to support life. However, a red blood cell will also read the part of the DNA that allows it to be pushed through tight capillaries while carrying hemoglobin, whereas a pancreas cell will instead read the instructions to sense blood glucose levels and secrete insulin if needed. Cells using only a subset of the available DNA information is why genetic diseases usually do not affect every cell. Even though a person may inherit a mutation from a parent that is then replicated in every cell of their body, the only cells that will show the disease are the ones that read that specific portion of DNA to perform their function. Cells that have not yet made the commitment to develop their specialized function are considered stem cells. Much effort has been

devoted to learning how to control what DNA stem cells read and thus the cell type they become in hopes of replacing or repairing malfunctioning cells.

This book will describe some of the unique qualities of 11 different cell types found in mammals while providing a broad overview of the basic biology of cells. The final two chapters will illustrate ways in which multiple cell types can coordinate, both for your benefit (the immune system) or to cause harm (cancer). It is my hope that readers appreciate this resource as a way to know themselves from a new perspective: as an amazing collection of distinct cells that accomplish remarkable feats, each of those cells built from nonliving materials that come alive to serve a specific role in forming you.

2

Red Blood Cells

In my view, red blood cells (RBCs) are a bit like decaffeinated coffee. Caffeine is a defining (and some might say essential) component of coffee. Similarly, RBCs are missing many defining features of cells. Among these features are *organelles*, subcellular structures that compartmentalize materials and serve distinctive roles within the cell. We will be examining the roles of many of these organelles in the context of other cell types in subsequent chapters. But for now, RBCs are a great place to start precisely because they are so relatively uncomplicated.

Just like a coffee bean, RBCs start out with all their parts, and it takes work to strip them down. They are born in the bone marrow, through a process called *erythropoiesis*. As part of their maturation, they begin to rid themselves of organelles. Most are broken down to the molecular level, but the largest organelle, the nucleus, is literally ejected from the cells of mammals. It is this *enucleated* version that is flowing throughout your body, and the absence of a nucleus is readily apparent as the center of RBCs is flattened and concaved as compared to its plump edges (Fig. 2.1). The nucleus contains a cell's deoxyribonucleic acid (DNA), and that DNA contains the genetic instructions for the RBC to build the features and functions that define it. Now lacking genetic information, an RBC cannot do much to influence its future. The nucleus contains a cell's its core role in your body is to simply be pushed about as it carries gases around. Intriguingly, other animals do not rid their RBCs of nuclei. The reason appears to be that mammalian RBCs are wider than our capillaries. In order to squeeze through some of these narrow passages (generally about 8 micrometers), the RBCs must bend, a feat that would be very difficult with a large, rigid nucleus. Given that the heart needs to push over 20 trillion RBCs around, I imagine there is also some benefit to losing the

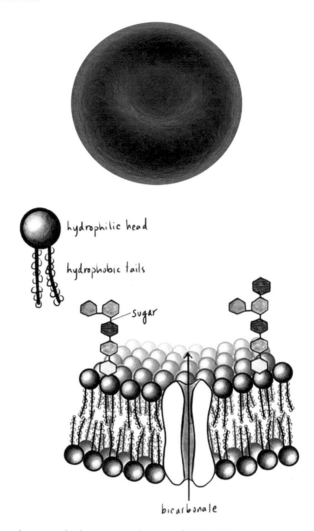

Fig. 2.1 *The shape and plasma membrane of RBCs*. The concave shape of an RBC reflects the loss of its nucleus (top, image from https://www.dreamstime.com/royalty-free-stock-photo-blood-cell-image1852625). The plasma membrane is what separates a cell from its environment (bottom). Two layers of phospholipids align tail to tail to form a barrier. Embedded in the plasma membrane are protein transporters to allow the passage of specific molecules, such as bicarbonate in RBCs. The outside of the membrane can be decorated with sugars; for RBCs, this designates ABO blood types

significant mass that a nucleus normally contributes. It might be helpful to imagine RBCs as round swim tubes floating along a lazy river with some very tight spots and with their absent nucleus as ousted human cargo.

Once released into the blood, an RBC survives about 120 days. During this time, it will travel through a human's circulatory system approximately once per minute. Every time it passes through the lungs, it will pick up oxygen (O_2) to deliver to all other cell types throughout the body. We all know that we need oxygen to breathe, but what does that mean at the cellular level? Well, the short answer is that O_2 serves as a final recipient in a process that allows cells to convert the organic molecules that we eat into usable energy. The details of that process will be revealed in Chap. 5. RBCs can also remove carbon dioxide (CO_2) – a molecule that carries no usable energy – from cells back to the lungs to be exhaled. The ability of RBCs to transport these two gases (O_2 and CO_2) is due to a molecule called *hemoglobin*, a protein produced in huge quantities by RBCs prior to losing their nucleus. Hemoglobin consists of four very similar parts called *subunits*, which can each bind one O_2 molecule. Hemoglobin requires iron to hold onto O_2, and iron deficiency in one's diet can lead to *anemia*, characterized by feeling tired, weak, and short of breath.

The ability to carry O_2 or CO_2 is not a simple process of direct exchange. CO_2 does not bind hemoglobin at the same site as O_2. However, when CO_2 does bind hemoglobin at its specific location, it alters the hemoglobin's shape in a way that encourages the release of O_2. In this way, RBCs have a way to sense the need for O_2. Regions of high CO_2 concentration reflect a need for more cellular energy, so as RBCs pass by, the O_2 needed to sustain energy levels is liberated. While O_2 and CO_2 do not bind hemoglobin at the same position, O_2 and carbon monoxide (CO) do. And unfortunately, CO is about 200 times better at binding hemoglobin than O_2, resulting in O_2 losing the competition most of the time. Exposure to high levels of CO prevents RBCs from delivering O_2 to cells throughout the body, halting their ability to convert energy and possibly causing organismal death. Thus, while RBCs have a relatively simple role in the body, it is certainly essential.

As indicated by their name, RBCs are red. Hemoglobin is responsible for the coloring, which can shift from dark to bright red as more O_2 binds (the appearance of having blue-colored blood in veins is an effect of different colors of light having a range of abilities to penetrate the skin). Notably, unlike RBCs, most mammalian cells are transparent and not free floating. The relative ease of access and the inherent color allowed individual RBCs to be seen using simple microscopes in 1658 (several years before Hooke conceived of the term "cell"). It took another 150 years for the detection of other types of blood cells, which generally require the addition of dyes. But importantly, the relative straightforwardness of both visualizing and counting blood cells allows for the precise measuring of diseases that manifest in this tissue, such as hemophilia, sickle cell anemia, and leukemia.

Like all cells, RBCs need a way of separating themselves from their environment, and this is accomplished by a *plasma membrane*. As the term "membrane" implies, it is a very thin structure: approximately 8 nanometers or just 1/1000th the width of an RBC. Its fundamental configuration consists of two layers of molecules called *phospholipids*. Individual phospholipids are often described as having two sides: one that is called the "head" and the other that consists of two "tails" (Fig. 2.1). These sides have a chemical feature that starkly distinguishes them and is central to their ability to form a barrier. The head domain consists of atoms that are attracted to H_2O, the most abundant molecule both within and in the environment surrounding cells. But the tail domains are repelled by water, a feature that is described as being *hydrophobic*. Because of this trait, the tail sides of phospholipids align with one another to avoid H_2O, while their heads face toward the H_2O-rich environments inside and outside of the cell. The double row of phospholipid tails creates a hydrophobic interior to the plasma membrane (as well as other intracellular membranes that we will encounter later) that is mostly impermeable to the molecules found on either side.

However, cells need to allow specific materials across this barrier. The ability to carefully regulate what is allowed in or out of a cell is accomplished by proteins that span the membrane. Proteins have a much more complex structure than phospholipids, which enables them to discern small chemical differences between the molecules they encounter. If it is a right match, the protein transporter can make temporary bonds with the molecule and provide it safe passage across the membrane. Some of these transport proteins will be found on most cells, for example, proteins that enable sugar to enter the cell. Other membrane proteins will be very specific to a particular cell type. For example, RBCs are just one of two cell types that pack their plasma membranes with a transporter called "Band 3," which lets a molecule called bicarbonate out of the cell (Fig. 2.1). Bicarbonate is essential to maintaining the blood pH at around 7.4.

Several other proteins as well as sugars can be found associated with the plasma membrane that are not involved in transport but instead function in communication. The communication can be very complex – for example, to receive external information, which initiates the internal processes of cell division or cell differentiation (something we will explore in the next chapter). Or the communication can be relatively simple, like a sign that conveys what type of cell it is. One set of molecules associated with the plasma membrane of RBCs that you are likely familiar with is that which defines blood type. Blood types "A," "B," and "O" are determined by short chains of sugars attached to the outer surface of the plasma membrane (Fig. 2.1). The sugars that constitute "O" chains are actually found in "A" and "B" chains as well, but "A" and "B" contain additional sugars that are unique to each. The function of these sugars within a person is not established, but they certainly play a role in cell recognition when transferred between people! If you receive blood from someone who has a different sugar decorating the surface of their RBCs than your own, your immune system will be able to recognize it as not part of your body and will mount an attack (how this is accomplished will be explained in Chap. 13). Because all the sugars that compose the chain of "O" are also found in "A" and "B" chains, "O" will not invoke an immune response in any recipient. Conversely, someone with "O" blood can only receive blood from an "O" donor because of the unique sugars found on RBCs with "A" and "B" chains. Blood types are also described as either positive (+) or negative (–); this has to do with a second molecule associated with the membrane of RBCs called the Rh factor. This is a protein that you either have (+) or don't (–). The Rh factor seems to be involved in transporting ammonium across membranes, but no physiological consequence of lacking it has been reported in 15% of people who are Rh-. Rh status must also be determined for successful blood donations; those who are negative will mount an immune response to RBCs containing the Rh protein. Because O- persons do not contain anything on the plasma membrane of their RBCs that can trigger an immune response in any recipient, they are termed universal donors.

Even though it serves as a barrier, the plasma membrane is not a stationary, solid structure. Rather, it is mobile and fluid (a cartoon video of its fluidity here: https://www.youtube.com/watch?v=LKN5sq5dtW4). It is able to maintain its integrity because of the drive of the hydrophobic nature of the phospholipid tails to remain together and away from H_2O. You have witnessed this force if you have ever mixed oil with water. Oil (be it olive or motor) molecules will come together in globs unless they are continuously and forcefully driven apart. The fluidity of the plasma membrane allows cells to be dynamic; membranes can be pushed forward as a cell crawls and can even form small, temporary appendages that can reach out to interact with other cells or grab onto stable, extracellular structures. These processes require an intracellular network called the *cytoskeleton*, which we will learn about more in Chaps. 6 and 12 when we encounter cell types that need to move in a purposeful way. The cytoskeleton is also responsible for cells having any shape other than being simply round, akin to an oil droplet. The concave nature of RBCs is maintained by components of the cytoskeleton, which also absorb physical stress while squeezing through capillaries, which might otherwise shear the plasma membrane. The correct cell shape, as determined by the cytoskeleton, matters. There is an inherited disease called spherocytosis, in which the cytoskeleton of RBCs does not assemble properly. This leads the RBCs to form a spherical instead of a flat, concave shape. The plasma membranes of the RBCs are easily damaged, resulting in cells living for only about 25 days instead of 120, leading to anemia.

RBCs are simple but very important cells. They are by far the most abundant cell type in humans; with between 20 and 30 trillion in an adult, they represent approximately 70% of total cells. You may be wondering – if the main role is to just be pushed around as they carry hemoglobin, why not just use "free" hemoglobin? Could hemoglobin not just travel via the liquid part (serum) of blood? In fact, RBCs do sometimes burst and release their hemoglobin (called *hemolysis*). However, at high levels, free hemoglobin damages tissues and is toxic to kidneys as they attempt to filter it out. So keeping hemoglobin safely packaged within RBCs is necessary.

3

Stem Cells

Throughout history, humans have sought out immortality through various means. Explorers have searched for the Fountain of Youth, and several religions promise eternity to the faithful. But it appears that all along, we have had the possibility of immortality within us via processes frequently applied by stem cells. Of course, one cannot substantiate anything as truly immortal, as long as tomorrow is still a possibility. However, stem cells use mechanisms that allow them to impede many of the cellular mechanisms that contribute to aging. As far as we can tell, some populations of stem cells can live indefinitely as long as they have a home – that home could be in a laboratory long after the mortal part of you is gone.

But we do not live forever, and at least up until now, that has not been the role of our stem cells. Stem cells serve as a repository of cells to provide the growth and maintenance of a multicellular organism. Importantly, they are flexible in terms of what type of cells their progenies become. Some are more flexible than others. The cells from very early embryos (prior to implantation) appear to be able to become any cell type that you could find in a corresponding adult. The term for this is *pluripotency*, and finding pluripotent cells that do not involve disrupting an embryo has been an intense avenue for research since the late 1990s.

Upon becoming an adult, a very small proportion of cells are set aside as stem cells and are much more limited in the types of cells that they can become. They generally can only contribute to one tissue or organ type. This ability is referred to as being *multipotent*. The first of these adult stem cells were found in our bloodstream and bone marrow and are called hematopoietic stem cells (HSCs). Including RBCs, which we just considered in the previous chapter, there are about a dozen distinctive cell types in our blood, and

L. Saucedo, *Getting to Know Your Cells*, https://doi.org/10.1007/978-3-031-30146-9_3

HSCs can produce them all. Their presence is why bone marrow transplants have been successful since the 1950s: after having his own leukemic HSCs destroyed with radiation, a patient received a bone marrow donation from his identical twin. This procedure earned the MD who performed it a Nobel Prize. Today, enough HSCs can be isolated from the blood circulating throughout the body, so that the much more invasive procedure to obtain bone marrow often is not necessary. The relative ease of obtaining immortal cells from blood eerily intersects with the myth of blood as a source of rejuvenation for Count Dracula.

More recently, adult stem cells have been found in many other locations, such as skin, brain, fat, gut, brain, and, intriguingly, the pulp of wisdom teeth. Throughout our life, these cells serve as the repository of replacement cells as older ones are lost or die. You may have heard previously that your skin is replaced every few weeks. This statement does not apply to all of your skin but does reflect the relatively quick turnover of the outermost layer, called the *epidermis*. The cells of the epidermis form a few dozen layers, and each cell spends about 2–4 weeks moving from the deepest layer, where they were first produced by stem cells, to the outermost layer, where they are sloughed off by daily activities. Importantly, when a stem cell divides, one of the two progeny remains a stem cell, while the other one is used to replenish the epidermis. In this way, the population of stem cells is maintained. The progeny of stem cells that do not remain stem cells has instead committed to further cell division and/or *differentiation*, which is the transformation to a specific cell type. In order to differentiate, a cell must begin to read or "turn on" the genes that provide specialized instructions to create the required structures and molecules. This process is generally initiated by interactions with the environment surrounding a cell. These interactions include contact with a variety of molecules that relay information about a cell's location as well as which other cells are immediate neighbors. A biological response such as turning on genes typically involves building (or deconstructing) biological structures. This work almost always involves *enzymes*, a group of molecules, usually proteins, that are able to break and form chemical bonds in a controlled manner. When it is time to use new genes, chemical signals are produced that enter the nucleus and direct an enzyme called *RNA polymerase (RNAP)* to "read" the appropriate genes. While genes are stored as dormant sequences of deoxyribonucleic acid (DNA), RNAP produces a complementary sequence in the form of a molecule called ribonucleic acid (RNA), which is actually put to use (details as to how are covered in Chap. 6). DNA and RNA are chemically similar, allowing the RNA to serve as a transcript of genetic information. However,

RNA is very unstable compared to DNA, which is important as that makes the directions of an RNA a short-term commitment. This enables cells to respond quickly when signals change and different instructions stored within DNA need to be utilized.

To learn about the process of stem cells producing differentiated cells in some detail, we will focus on intestinal stem cells (ISCs), whose location and role in tissue maintenance are relatively easy to visualize. The small intestine needs to absorb food but also secretes mucus, hormones, and chemicals that regulate the population of bacteria living in your gut. ISCs produce the different cell types that accomplish these roles. The surface that faces the interior (or *lumen*) of the small intestine is packed full of projections called *villi*. This greatly increases the surface area for both absorption and secretion. Between the villi are invaginations into the tissue surface called *crypts*. ISCs are located in the base of the crypts, generally restricted to what is referred to as the +4 position and below (Fig. 3.1). When an ISC divides, one of its progenies will eventually commit to becoming one of four cell types, while the other remains a stem cell. One of the possible cell types is called a Paneth cell – it will remain in the base of the crypt and differentiate into a cell that secretes molecules that counter the overgrowth of bacteria. The other three cell types are produced by ISCs at the +4 position and will be pushed up toward the tip of the villi, spending only 5 days doing their specific jobs before they are shed into the lumen of your gut to become part of a bowel movement (a video depicting a cartoon version of this process can be seen here: https://www.youtube.com/watch?v=qq5k1sWqLO0). Most of these cells will differentiate to become enterocytes, the cells that absorb nutrients. A few will instead differentiate into goblet cells, which secrete mucus to provide a coating to help move things along the intestinal tract. And finally, a few will commit to becoming enteroendocrine cells, whose job is to secrete hormones that help control digestion and appetite. Each of these final cell fates is determined by a combination of signals from other cells and the environment. Importantly, cells that remain stem cells are also dependent on interactions with nearby cells and the local environment, which together are termed the *stem cell niche*. For ISCs, the niche involves cells found underneath the surface of the crypt. When the ISC divides, one progeny maintains interactions with the niche to remain a stem cell, while the other cell loses those interactions and gains new ones to promote continued division and differentiation (Fig. 3.1). Holding onto their niches and avoiding differentiation help ensure stem cells remain available to continue providing new cells for an organism's lifetime.

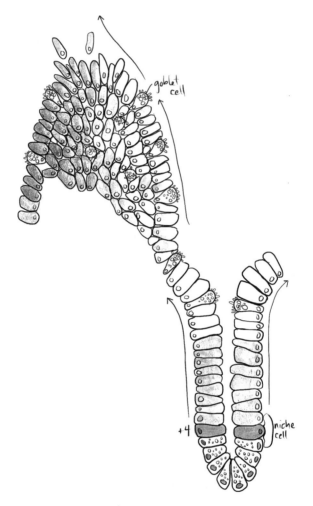

Fig. 3.1 *The intestinal crypt as a model for tissue maintenance by adult stem cells.* Stem cells are found near the base of a crypt and produce four cell types. Stem cells located at position demarcated +4 produce cells that are pushed up the microvillus and eventually shed into the gut. Stem cell populations are maintained by interactions with neighboring niche cells

Importantly, stem cells have many other features that keep them from aging. Aging at the cellular level is primarily defined by the condition of a cell's DNA. The life experience of a cell can be revealed by its DNA in a couple of ways. First, there is a mechanism for DNA that serves as a means to measure how many times a cell has divided. Each cell division requires a duplication of its DNA content so that the progeny receives its instructions. Organisms package their DNA as *chromosomes*, and the enzyme with the capability of making copies of an organism's DNA is called *DNA polymerase*.

However, DNA polymerase is unable to replicate some of the DNA at the very ends of chromosomes that are linear (like ours). And that means our chromosomes get slightly shorter with each cell division. Generally, after dozens of cell divisions, the ends of chromosomes (called *telomeres*) become short enough to alert the cell to cease further division and enter a state that is called *cellular senescence*. This is a protective mechanism that happens prior to losing DNA that actually encodes for genes. However, some stem cells have the ability to counter the shortening of their chromosome; they harbor an enzyme called *telomerase*, which adds DNA back to telomeres.

Telomerase is encoded in the DNA of all our cells and is "on" at its highest levels in embryonic stem cells. As organisms develop into adults, telomerase is primarily turned "off" in cells, with a few exceptions, including some adult stem cells. The power of telomerase to extend the lifespan of cells was demonstrated over 20 years when researchers manipulated differentiated cells growing in tissue culture to express telomerase. These cells were able to divide at least 20 more times than normal in the laboratory setting. Telomerase has garnered the interest of humans hoping to extend their lifespan with numerous supplements available purporting to activate its function. However, cellular senescence exists for a reason: cells experience and accumulate damage over time. This damage can lead to a cell becoming cancerous (discussed in Chap. 14). Cancerous cells cannot do much damage if they are unable to divide and create more of themselves, and thus those cancers we do detect almost always have successfully turned on their telomerase gene.

Another way that aging can usually be revealed is through the accumulation of mutations in the cell's DNA. DNA mutations can occur through numerous mechanisms, and stem cells proactively counter them. One example is that stem cells use metabolic pathways that cause less damage. Most cells in our body undergo what is termed *aerobic respiration*; it is the most powerful way to convert food into usable cellular energy with the help of oxygen (to be described in Chap. 5). However, it also produces molecules called *reactive oxygen species (ROS)*, which alter and impair the molecules they encounter – including DNA. If you have ever bought a product because it contains antioxidants, the benefit you hope to reap is to counteract the damage wrought by ROS. Stem cells minimize their use of aerobic respiration, instead relying on its relatively inefficient counterpart, anaerobic respiration, for energy conversion while protecting their DNA. However, mutations are a natural event every time a cell replicates. DNA polymerase inherently makes mistakes as it attempts to double genetic information. Although DNA polymerase has the ability to proofread its work and additional enzymes in our cells are on the constant lookout for errors, a handful of mutations will become permanent.

This imperfect process is important for evolution as diversity is ultimately dictated by variations in DNA sequence. Stem cells minimize the accumulation of these mutations by sequestering one of the cells immediately after one cell division (and the replication of DNA) to remain a stem cell, while the other cell goes on to proliferate numerous times (Fig. 3.1).

The rise in knowledge about stem cells has energized the field of regenerative medicine, which is centered on replacing tissues and organs damaged by aging or injury. Mammals have limited regenerative capabilities, with the liver as a distinctive organ, which can regrow on its own and function normally following the removal of up to three fourths of its original mass. This feat is accomplished because most liver cells can "dedifferentiate" back toward an uncommitted stem cell. Could this outcome be replicated for other, even all, of our organs? Finding such flexible cells in adults is a very active area of research, but technology that can coax adult cells "back" toward pluripotency already exists. In a process that creates *induced pluripotent stem cells* (iPSCs), genetic engineering is used in tissue culture to turn on a handful of genes that are normally dormant in skin and blood cells. This entices them to dedifferentiate themselves toward an embryonic-like state. Because these cells have been genetically manipulated, there are safety concerns about using them for treatment. Also, some of the genes used for this process are also known to partially contribute to some cancers. Finally, the process of directing the cells to correctly differentiate upon reintroduction into patients cannot be fully controlled. Nonetheless, several clinical trials using iPSC approaches have recently started.

In recent years, the idea of storing cells found in the blood of umbilical cords of newborns as a personal bank of stem cells for future therapeutic use has rapidly gained momentum. While not containing any pluripotent cells, cord blood contains a multitude of different, multipotent stem cells. Already approved for treating diseases arising from blood, such as leukemias, there have also been cases reporting the success of the use of cord stem cells to treat cerebral palsy. Some studies using cord stem cells for spinal and cardiac repair have also shown promise.

The strength of using a patient's own stem cells for treatment is that it averts an immune reaction. But if the health issue arises from an inherited genetic condition, it will also be present in their stem cells. Recent advances in molecular biology to repair or replace genes may provide a path to solving this problem. While numerous hurdles remain, the ability to eventually exploit stem cells to rejuvenate aging organs seems feasible and may help eliminate death by "old age," an essential hurdle to overcome in order to achieve immortality.

4

Epithelial Cells

This cell type is likely the one that you have the most experience seeing, comprising the surface of your skin and sometimes floating free as dust particles crossing a ray of sun.

One reason that epithelial cells are relatively effortless to encounter is that this cell type often has one side in contact with the environment. They are most obvious on our outer surface functioning as skin, but they are also in direct contact with the environment that you allow into your body, like air into your lungs or food and drink into your digestive system. An example is the intestinal enterocytes that absorb food, which we encountered in the previous chapter. Therefore, these cells are on the front line of both physical protection and decision-making regarding what to let pass into underlying tissues. Many epithelial cells also have the ability to secrete compounds. In some cases, these secretions contribute to protection, such as mucus to capture bacteria in your lungs and intestines. Other epithelial cells have a primary role to secrete materials specific to the gland they call home, for example, releasing sweat or saliva. Some derivatives of epithelial cells lose direct contact with the environment during embryonic development and become highly specialized. Two examples are hepatocytes and endocrine cells, which will be discussed in subsequent chapters.

Epithelial cells come in a variety of shapes and are often grouped by that measure (Fig. 4.1). Relatively flat cells are called *squamous*, whereas epithelial cells with more three-dimensional (3D) proportions are called *cuboidal* or *columnar*. Epithelial cells can function as a single layer (*simple epithelial*) or can pile on top of one another (*stratified epithelial*). In general, layers of epithelial cells offer more protection: four to five layers of stratified squamous epithelial cells serve as the outermost surface of our skin.

L. Saucedo, *Getting to Know Your Cells*, https://doi.org/10.1007/978-3-031-30146-9_4

Types of Epithelium

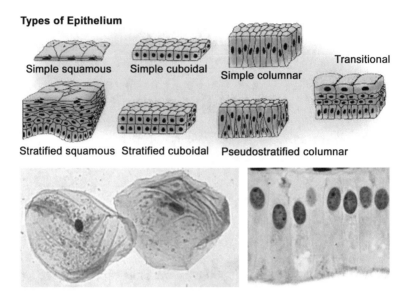

Fig. 4.1 *Types of epithelial cells*. Cartoon depictions (top, from https://en.wikipedia. org/wiki/Epithelium#/media/File:Illu_epithelium.jpg) and microscopy images of epithelial cells (bottom left: squamous cheek cells (Adapted from Getty Image), bottom right: simple columnar intestinal cells (https://www.carolina.com/histology-microscope-slides/human-simple-columnar-epithelium-slide-7-m-he/312426.pr))

Being in direct contact with the environment often means experiencing physical pressure. Consider each time you bump into something or the contractions needed to propel food through your digestive tract. In order to maintain integrity, epithelial cells need to be able to absorb that physical stress. One way cells do this is via the cytoskeleton, a network of proteins found inside cells. Similar to an organism's skeleton, the cytoskeleton is integral to establishing overall shape as well as absorbing impact.

There are different components that contribute to the cytoskeleton, some that are very dynamic and which will be discussed in future chapters. But the components of the cytoskeleton that does the bulk of the work of absorbing stress are called *intermediate filaments*. They are given this name because of their in-between size compared to other members of the cytoskeleton. In epithelial cells, intermediate filaments are constructed from a protein called *keratin*. It is likely that you have heard of this protein previously as it is found abundantly outside of cells as our hair and nails. Within cells, intermediate filaments form a relatively simple and stable scaffolding. Importantly, the scaffolding of one cell is bound indirectly to the scaffolding of neighboring cells and to other extracellular scaffolding proteins so that each cell is not alone in absorbing stress (Fig. 4.2).

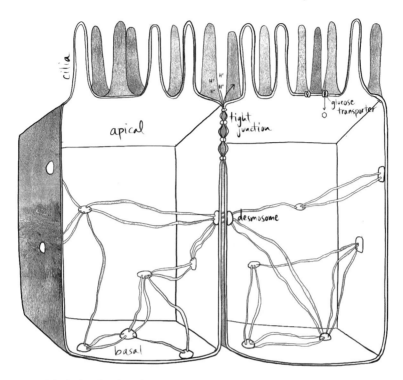

Fig. 4.2 *Epithelial cells absorb stress and control what passes to underlying tissue.* Intermediate filament scaffolding within and between epithelial cells and underlying tissue serves to counter mechanical disruption. The space between epithelial cells is sealed by tight junctions, impeding the passage of even the smallest atom (H⁺) via *paracellular* transport unless warranted. Most transport occurs transcellularly, via proteins embedded in the membrane, such as the glucose transporter. *Cilia* increase the surface area for transport (both into and out of the cell) and can also sweep materials along the apical side of cells

However, intense or persistent pressure can break this scaffolding. Although other adhesions between cells can maintain their attachment to one another, whole sheets of cells can be released from underlying tissue if the intermediate filament structure is disrupted. This creates a space that is recognized as a wound, filling up with liquid to form a blister. The critical importance of epithelial cells' ability to absorb mechanical stress was dramatically revealed when mice were genetically modified to contain a mutant version of keratin. The stress of moving through the birth canal caused the *epidermis* (the part of the skin consisting of epithelial cells) to slough off. Additional work in mice helped pinpoint the cause of a human disease characterized by easy blistering and skin loss, called *epidermolysis bullosa simplex (EBS)*, now confirmed to be due to inherited genetic mutations in keratin proteins. There is quite a range in how severe EBS manifests in patients, but skin infections are always a concern (Fig. 4.3).

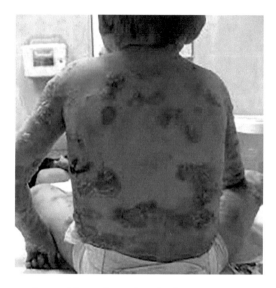

Fig. 4.3 *A young patient with epidermolysis bullosa simplex.* (From https://escholarship.org/content/qt01h2c4k6/4.jpg)

As the frontline barrier to the environment, epithelial cells need to prevent unwanted materials from gaining access to the underlying tissue. Therefore, they are tightly packed together. But to truly seal the space between cells, they build formations called *tight junctions*. These junctions are formed by the interaction of proteins embedded in the membranes of neighboring cells. They are able to block the passage of H_2O and even the smallest atom (H^+) (Fig. 4.2). However, tight junctions can also selectively allow movement not only of atoms but also of entire cells! For example, tight junctions between epithelial cells in the kidney allow calcium (Ca^{2+}) and magnesium (Mg^{2+}) to reenter the bloodstream instead of moving onto the bladder to be lost as waste. Cells of your immune system are also able to coax open tight junctions as they proactively search for invaders. A specific example is dendritic cells, which send "arms" between the gut epithelium to "sample" the bacteria in your digestive tract; we will see what they do with that information in Chap. 13.

Impressively, the dendritic cells contain some of the same proteins on their surface that are used in tight junctions, helping to keep the barrier intact even as they insert themselves between epithelial cells. The importance of tight junction function has become more evident in the past couple of decades as research into inflammatory bowel diseases (IBS) shows a breakdown of the integrity of the epithelial barrier. In a sense, this disruption is perceived by the immune system as a persistent wound, leading to irritation and hindering digestion.

In addition to having tight junctions control what substances are allowed to move through the space between cells (*paracellular transport*), epithelial cells have proteins on their membranes that bind to and allow the entry of molecules into the cell itself (*transcellular transport*, Fig. 4.2). One specific example is a protein found on the epithelial cells of the gut that brings a sugar called glucose into the cell. This protein is found in high concentrations on the *apical* side of the epithelial cells, which faces the lumen of your intestines (Fig. 4.2). As you digest your food, individual glucose molecules are released and bound in a specific manner to the transport protein, which then provides safe passage for the glucose molecule from the membrane, which would otherwise exclude it. Because we want to bring in more than just half of the sugar we digest, instead of relying on simple diffusion, this transport protein uses energy to concentrate glucose into the gut epithelial cells. Once in the cells, the glucose then moves out of the cell into the bloodstream via a different protein transporter, found on the other side of the epithelial cell, called the *basal* side. The transport protein that performs this role does not use energy and allows glucose to move down its chemical gradient, resulting in a blood glucose concentration of ~80 mg/dL. This level then serves as a healthy supply of glucose for all other cells in the body to perform their functions. Tests for blood sugars are measuring the concentration of glucose. If it remains high (over 125 mg/dL) even while fasting, that is indicative of diabetes, a condition we will touch on in Chap. 10.

Some epithelial cells specialize in secreting molecules out of them (that process will be described in some detail in Chap. 6). This can be for the purpose of protection, usually in the form of mucus. Mucus contains sugars and proteins, which are ejected via secretion from the apical surface, the cell side in contact with the environment. This viscous material protects in two ways: 1) it coats cells and helps absorb mechanical stress, and 2) it traps components from the environment, including bacteria and viruses. Importantly, trapping materials is not enough; over time, they would build up to levels that would interfere with tissue function. The mucus (and whatever it has trapped) needs to be cleared away. In the respiratory tract, epithelial cells play an essential role in "sweeping" mucus up toward the throat. This requires the cells to have *cilia*, which look like small, cellular appendages extending from the apical side (Fig. 4.2). Within each cilium is a complex cytoskeletal structure consisting of *microtubules* (discussed in depth in Chap. 12), which allow the cilia to sway in a coordinated manner. Unicellular organisms can use cilia to swim, but since epithelial cells are fixed in place, the result is that the mucus layer they are in contact with gets pushed. So the mucus is carried up your trachea and back into your throat to be swallowed unconsciously a couple of times a minute, resulting in digesting over a quart a day!

In spite of all the protections, being on the front line with the environment means epithelial cells do take on a lot of damage. Therefore, they are replaced quickly. Intestinal epithelium regenerates especially rapidly, with a cellular lifespan of only about 5 days. It has been estimated that 1 gram of human fecal matter contains ten million epithelial cells that have been shed from the gut! While damaged epithelial cells may not be able to perform their normal functions properly, a bigger concern is that environmental insults may lead to genetic changes that empower epithelial cells rather than disable them. This type of damage contributes to the progression of cancer and likely reflects why the vast majority of cancers arise from epithelial cells.

5

Hepatocytes

If you have ever thought about the function of your liver, it is likely that it was either in the context of its role in removing toxins or its susceptibility to damage by high levels of alcohol or certain medicines. These two common impressions about the liver are related to one of its key functions: to filter our blood. During that process, it encounters molecules and can alter their composition. Intriguingly, while this alteration can successfully prevent the accumulation of certain harmful molecules in our body (i.e., "detoxification"), it sometimes creates a more toxic substance than it first encountered, which can then damage the liver and other organs. The liver is also a central hub of energy conversion, also known as *metabolism*. In this role, it still alters molecules, but as a means to regulate the balance of stored vs available fats, carbohydrates, and proteins. If you are a fan of chocolate bars, you can thank your liver for preventing them from poisoning you and also implicate it in any consequent weight gain!

The vast majority of cells that make up the liver are hepatocytes. They are a type of cell that contains an unusual amount of an intracellular structure called the *smooth endoplasmic reticulum* (SER) (Fig. 5.1). The SER consists of a membrane that forms folds and tubes. The structure of the SER membrane shares some properties with the plasma membrane, as described in Chap. 2, using two layers of phospholipids to create a barrier between water-abundant compartments. Only in this case is the barrier formed between the cytoplasm of the cell and the interior space, or lumen, of the SER. The abundance of SER in hepatocytes reflects its central part in both detoxification and metabolism; in fact, the surface area of the SER is approximately eight times as much as the plasma membrane in these cells.

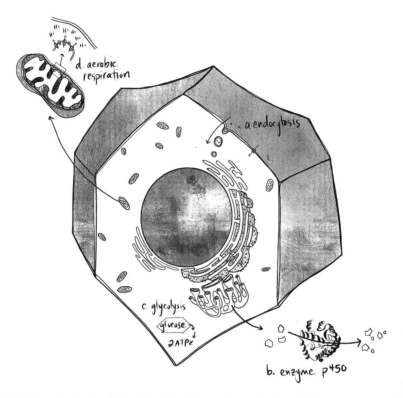

Fig. 5.1 *Hepatocyte's role in detoxification and metabolism.* A hepatocyte can bring in materials that we have digested via endocytosis (**a**) Some of those materials will be acted upon by p450 enzymes in the smooth endoplasmic reticulum (SER), which attaches oxygen (O) to molecules in order to help clear them from our bodies (**b**) Other molecules will be shuttled into metabolic pathways, like glycolysis in the cytoplasm (**c**) and aerobic respiration in mitochondria (**d**) These two processes transfer energy from the chemical bonds in food that we eat (like glucose) to a molecule called ATP

Material that we decide to eat or drink can enter our bloodstream by way of our intestines. But before circulating throughout our body, they are carried in the blood to the liver. Here, capillaries containing small pores allow blood to leak out and saturate nearby hepatocytes. Hepatocytes are then able to sample components in the blood via *endocytosis*, a process by which the plasma membrane invaginates and pinches to form intracellular vesicles containing extracellular substances (Fig. 5.1). Additionally, hepatocytes contain numerous transporters to allow direct passage across their plasma membrane via transcellular transport, as we saw in the previous chapter. Once inside the hepatocyte, molecules can encounter enzymes that are attached to the membrane of the SER, called p450s (Fig. 5.1). Humans have at least 57 versions of p450s, about half of which perform reactions that are meant to detoxify. In most cases, detoxification means that the p450 adds oxygen to a molecule that is not recognized as useful to an organism; generally, these are molecules that were not needed for our ancestors to survive. This helps ensure that the unfamiliar molecule is able to dissolve in water and be quickly excreted (via urine) rather than allowed to build up in our bodies and possibly cause harm. Humans have p450s, which efficiently act on compounds in chocolate that could otherwise poison us... unfortunately, dogs do not.

Some molecules that humans purposefully ingest that are targeted by p450s are medicines. P450s act on some antibiotics, birth control, and painkillers. And how p450s interact with one drug can alter their function on another. For example, taking a certain class of antibiotics accelerates how quickly p450s break down estrogen, which likely reduces the effectiveness of hormonal birth control. Variations in the specific versions of p450s that a person has can alter the efficacy of the medications they take, and this is currently an active area of investigation in uncovering why some patients are more or less responsive to standard doses of pain medication. For example, genetic variation among p450s correlates to a 30-fold difference in the rate codeine is broken down in patients. Further investigation into the role of p450s in detoxification will likely help personalize drug regimens and dosing to a person's genetics.

Strikingly, p450s' modification of chemicals sometimes makes the chemical more dangerous. This is the case with many chemicals that are considered *carcinogens*, substances that increase the onset of cancer. For example, the chemical benzo[*a*]pyrene is produced when something organic is burned, including tobacco and grilled food. After encountering and being modified by p450s, benzo[a]pyrene gains the ability to interact with deoxyribonucleic acid (DNA), which can lead to genetic mutations that promote cancer (discussed more in Chap. 14). Similar to sensitivity to pain medications, variations in the p450s that individuals have appears to correlate to susceptibility to cancers involving chemicals processed by the liver.

Following the digestion of food, the building blocks of fats, carbohydrates, and proteins are first encountered by hepatocytes. Here, a decision of whether to store or release these building blocks is made. This choice reflects the two arms of metabolism: *anabolism*, which supports the storage of digested materials, and *catabolism*, which promotes the immediate breakdown of these materials. To explore metabolism in some detail, this chapter will focus on carbohydrates. A common building block of carbohydrates is the simple sugar, *glucose*. The amount of glucose in the blood is monitored and regulated by hormones released by the pancreas (to be discussed in Chap. 9). One of those hormones, *insulin*, increases when blood glucose levels are high, and another hormone, called *glucagon*, is increased when blood glucose levels are too low. Due to its proximity to the pancreas, the liver is the first organ to be exposed to these hormones, and the plasma membrane of hepatocytes is studded with receptors for both. After a meal, the blood levels of glucose rise and insulin is released. In response, hepatocytes take in glucose via a transport protein and attach glucose molecules together to form long chains called glycogen. This is the storage form of carbohydrates in animals (as opposed to plants that store glucose as starch). When blood glucose levels drop, glucagon levels rise to signal that it is time to release glucose back into the bloodstream. Glucose is needed by all cells as an energy source; in most cases, this will be accomplished by a process called aerobic respiration (mentioned briefly in the previous chapter), which is a series of steps to convert food into a molecule called adenosine triphosphate (ATP). In hepatocytes, ATP is used for all sorts of cellular processes, including the conversion of digested food into its storage form, as described above.

Unlike glucose, or other molecules that we consume for energy, ATP is a relatively unstable molecule. This is beneficial because it means it is relatively easy for it to transfer energy. This energy transfer can be used to chemically alter another molecule or to physically move molecules. While the term "burning calories" might make sense because exercise causes you to sweat, it is essential that most of the energy transfer needed to support life does not manifest as heat; otherwise, we would truly ignite. Therefore, energy is extracted from glucose slowly via many, many steps to reduce heat production and create a pool of ATP. If oxygen is present and cells contain mitochondria, energy from a single glucose molecule can produce over 30 ATPs.

When it is time to extract energy from glucose, it enters a metabolic pathway in the cytoplasm, called glycolysis (Fig. 5.1). Over a series of ten chemical reactions, only two ATPs are produced. The rest of the energy remains in other products produced, called pyruvate and nicotinamide adenine dinucleotide (NAD) + hydrogen (NADH). For the rest of aerobic respiration, we need to move into the mitochondria. The mitochondria are cellular organelles that contain two membranes – an outer one, which is in contact with the cytoplasm, and an inner membrane, which is highly convoluted and creates two compartments within the mitochondria – the intermembrane space and the matrix (Fig. 5.1). Pyruvate is transported across both membranes into the matrix, where it will enter another metabolic pathway called the tricarboxylic acid (TCA) cycle. This pathway will take additional eight steps to convert most of pyruvate's energy into more NADH molecules. NADH is key in the production of the bulk of ATP. This is because NADH is able to accept and donate electrons, which is the source of energy in molecules. NADH was produced during glycolysis and the TCA cycle because it took electrons away from glucose and pyruvate. It is now ready to donate those electrons to an arrangement of molecules embedded in the inner membrane of the mitochondria, called the electron transport chain (ETC). The components of the ETC take turns in accepting and donating electrons in a directed manner, with the electrons losing energy along the way. The energy removed from electrons as they travel through the ETC does not result in ATP directly. Instead, the energy is used to move protons (H^+) across the inner membrane, concentrating them in the intermembrane space (Fig. 5.1). This unequal distribution of protons is a form of potential energy, similar to a dam creating a reservoir to store water. A final player in aerobic respiration is a complex called ATP synthase. It is also embedded within the inner membrane. It contains a channel to allow protons to leave the intermembrane space and return to the matrix. When this happens, the potential energy of the proton gradient is converted to mechanical energy, causing parts of ATP synthase to move like a rotor (see an animation of the ETC, proton gradient, and ATP synthase working together here: https://www.youtube.com/watch?v=39HTpUG1MwQ). This physical movement allows ATP synthase to clamp down on the breakdown products of used ATP and reform it. ATP is released, and the rotor continues, producing more. Meanwhile, as electrons reach the end of the ETC, they join with oxygen to form water. This last step is why we need oxygen. If oxygen is not available to remove electrons at the end of the ETC, everything gets backed up, no new electrons can be accepted, and the ETC halts. No new protons can drive the rotor of ATPase, preventing the synthesis of the vast majority of ATP. Within just a few minutes, cells run out of energy to function and support life.

These dozens of small steps are needed to convert energy in a way that minimizes the loss of energy as heat. However, sometimes it is beneficial to allow heat to be produced during catabolism. One way this is accomplished is via molecules that can be found on the inner membrane of mitochondria. These molecules are called uncoupling proteins (UCPs). Like ATP synthase, they also allow protons to move across the inner membrane of mitochondria. However, this relocation does not result in any new chemical bonds, and the potential energy is instead converted to heat. Thus, UCPs literally uncouple the proton gradient from ATP synthesis. Uncoupling proteins are abundant in some *adipocyte* cells, called brown fat. This type of fat cell is relatively abundant in infants, helping to regulate the body temperature of this vulnerable population. Some brown fat cells remain in adults, and while UCP activity does inversely correlate with weight gain over time, it does not appear to contribute to obesity compared to other factors. However, ingestible molecules that imitate the function of UCPs exist. One such molecule, 2,4-dintrophenol (DNP), was originally available in the 1930s for weight loss since the subsequent reduction in ATP limits the activity of anabolic pathways to store food (in either fat or muscle). But inadequate levels of ATP and overheating led to severe side effects and even fatalities. Unfortunately, in spite of being banned in the United States soon after, there have been several waves of illegal DNP use. The development of less toxic versions of UCP-like pharmaceuticals is an ongoing field of research to combat obesity.

If you are feeling overweight, it is likely you are thinking about fat and not your liver. However, the hepatocytes are the primary site of converting carbohydrates into fats. These fats are then packaged along with cholesterol (also produced in the liver) into particles called *lipoproteins* and released into the bloodstream. While both cholesterol and fats are necessary for all cells, excess will be stored in adipocytes, which, in turn, have the ability to grow substantially in size, making them far more conspicuous than the liver. The failure of the liver to send excess fats to adipocytes leads to fatty liver disease. While relatively common and asymptomatic, fatty liver disease can progress and cause severe damage.

As mentioned in the previous chapter, the liver is unique in its ability to regenerate, which may be a consequence of the substantial damage it incurs while its functioning in detoxification. However, its regenerative capacity leads to misery in Greek myths. Both Prometheus and Tityus are tortured by having their livers eaten by eagles and vultures, respectively, only to have the organ renewed for eternity.

6

Fibroblast Cells

This cell type might resonate with people who consider themselves loners. Fibroblasts often prefer to not be in close contact with other cells. These cells can accomplish self-isolation in part by having the ability to secrete materials, collectively called the *extracellular matrix (ECM)*, into their surroundings. This matrix consisting of proteins and sugars physically excludes cells from occupying the same space. If an area does get too crowded, fibroblasts are highly adept at moving to a new location. Fibroblasts are the cells found predominantly in connective tissues – which include bone and fat. They are also essential in the process of wound healing.

Fibroblast cells can take on different shapes, depending on the connective tissue they inhabit. In this chapter, we will focus primarily on fibroblasts found in the *dermis* layer of your skin. The dermis is underneath the epidermis, which, as we learned in Chap. 4, consists of tightly packed epithelial cells. A barrier, called the basement membrane, separates the two very different compartments of the skin. Dermal fibroblasts are generally fairly flat, and each can form a unique shape that gives the impression that it is stretching out its plasma membrane in multiple directions to reach something (Fig. 6.1). The extensions help the cell to secure itself to its environment and also to detect the unwanted presence of nearby cells.

The cell shape and the ability to move around via crawling are both achieved predominantly by a component of the cytoskeleton referred to as *microfilaments* (Fig. 6.1). Microfilaments get their name by being the smallest element of the cytoskeleton. They are also able to rapidly grow and shrink, form crosslinks, and bundle together. Importantly, because of their small size and ability to grow, they can gently push against a plasma membrane to move a cell in

© The Author(s), under exclusive license to Springer Nature Switzerland AG 2023
L. Saucedo, *Getting to Know Your Cells*, https://doi.org/10.1007/978-3-031-30146-9_6

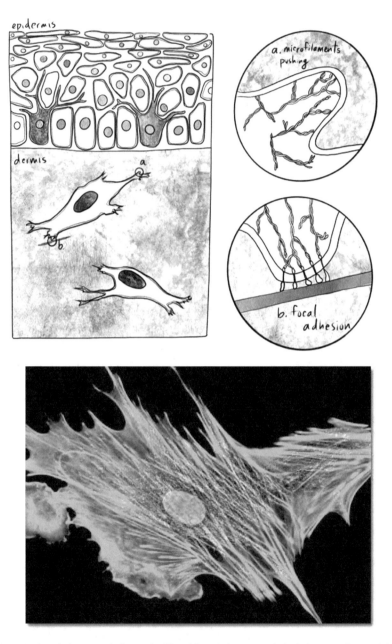

Fig. 6.1 *Role of the cytoskeleton in fibroblasts' shape and their interactions with the environment.* The top illustration is a side view of the skin with fibroblasts in the dermis layer using microfilaments to extend the plasma membrane (**a**) or to secure themselves to their environment with focal adhesions (**b**). In the bottom image, fluorescence microscopy is used to visualize microfilaments in green (with the nucleus in blue and mitochondria in red), from https://micro.magnet.fsu.edu/primer/techniques/fluorescence/gallery/cells/apm/apmcellsexlarge4.html

that direction without disrupting the barrier function of the membrane. I tend to think of microfilaments as fingers coaxing the shape of bowl on a pottery wheel as opposed to using other components of the cytoskeleton that are more analogous to trying to shape a pot with your fist.

Microfilaments are also able to bind to proteins in the plasma membrane, called *integrins*, which are simultaneously binding to static structures outside the cell in the ECM, allowing a fibroblast to temporarily grip onto its environment as it moves. Though not exactly speedy at about 1 micrometer/minute, fibroblasts enjoy a lot more freedom than most cells that reside in solid tissues (a beautiful time lapse of a fibroblast crawling through ECM in a tissue culture plate can be seen here: https://www.nidcr.nih.gov/news-events/nidcr-news/2021/creatures-crawling-within). When a fibroblast has found a space wherein it wants to remain, it creates long-term complexes between the cytoskeleton and the environment, called *focal adhesions* (Fig. 6.1).

While many cell types can secrete extracellular matrix, fibroblasts are especially proficient. You have likely heard of one of the most abundant molecules that they secrete: *collagen*. Collagen makes up about 30% of the protein found in humans and is very predominant in bone and skin, serving the role of providing strength and pliability. In skin, collagen is concentrated in the dermis layer because that is where the fibroblasts reside (Fig. 6.1). Many skin products contain collagen, with the promise of improving skin elasticity. However, the ability of collagen to penetrate the epidermis to reach the dermis is very unlikely due to the tight junctions previously discussed in Chap. 4. In order to manually increase collagen levels, it must be injected below the epidermis via a needle. This will only be temporary, however, as fibroblasts also produce enzymes to break down the collagen and other ECM components for continuous renewal.

The process of secretion begins inside a cell's cytoplasm as a particle called *ribosome* converts genetic information into a protein. The ribosome is able to read the language of a ribonucleic acid (RNA) molecule that was formed by copying information from a gene encoded by deoxyribonucleic acid (DNA) in the nucleus (a function carried out by RNA polymerase, as described in Chap. 3). Recall from the introduction that each cell type employs only the genes it needs in order to perform its function. This restriction is primarily regulated by producing RNA messages only from the required genes. Ribosomes cannot read DNA directly, and only RNA is allowed to exit the nucleus to reach the ribosomes in the cytoplasm. RNA is a strand of *nucleotides*, where groups of three consecutive nucleotides can represent one amino acid. While we have 20 amino acids to use to make proteins, collagen is a relatively simple protein consisting of an unusual amount of just two amino acids, called glycine and proline. As the ribosome converts an RNA message to a collagen protein, it pushes the collagen into a subcellular compartment called the *rough endoplasmic reticulum* (*RER*, Fig. 6.2). Suspiciously similar sounding to the smooth endoplasmic reticulum (SER) that we learned about in the preceding chapter, the difference is due to the density of ribosomes attached, which gives the RER a bumpy appearance under the electron microscope. The RER is surrounded by its own membrane, and the collagen is deposited to the inside, or *lumen*, of the endoplasmic reticulum (ER). Proteins that have entered the ER can now continue along the secretory pathway of a cell, which means the proteins can move to another membrane-bound set of organelles called the *Golgi apparatus.* Many types of chemical modifications can be made in the Golgi, but the most common is the addition of small carbohydrates to the much larger proteins. From there, proteins can move via small vesicles to the plasma membrane, where they can either remain or, like collagen, be ejected into the extracellular space via a process called *exocytosis.*

Collagen is a simple but oversized protein. While in the cell, it consists of three strands that twist together in a *helix* shape (Fig. 6.2). Once secreted, multiple sets of these triple helices come together lengthwise and also align alongside one another via bonds called cross-links to form bundles that can be as wide as a cell. The cross-linking of individual collagens into larger structures outside of the cell is essential to fulfilling collagen's functions. Too little cross-linking is implicated in Ehlers-Danlos syndrome (EDS), which is characterized by hyperflexible joints and loose skin. On the other hand, too much cross-linking is seen as people age and leads to brittle bones and contributes to inflexible skin (aka wrinkles).

Fibroblasts are able to receive chemical signals that stimulate the secretion of materials needed to build an extracellular environment appropriate to the region in which they reside. These tissues end up consisting of mostly extracellular matrix molecules with fibroblasts scattered throughout. Once a tissue is established during development, fibroblasts begin a new role of tissue maintenance and repair. Their role in wound healing is especially remarkable. When the tight connections between epithelial cells in the epidermis are broken, fibroblasts in the underlying dermis layer respond by proliferating and migrating toward the injury, a process that could literally be described as your skin crawling. They are able to track a chemical trail to the wound provided by the first responders of the immune system using receptors on their plasma membrane. Once bound to the chemical cue, the receptors can recruit microfilaments within the fibroblast to assemble at the same location and propel the

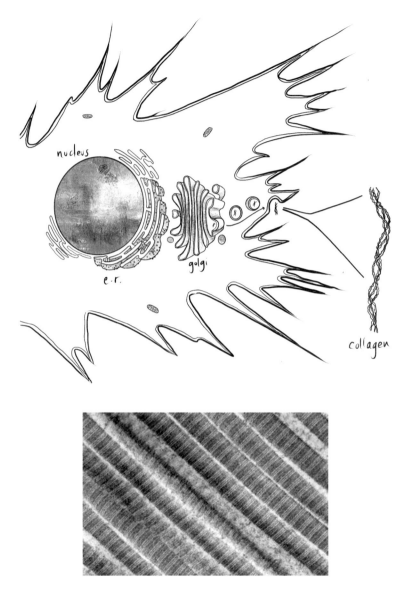

Fig. 6.2 *The secretory process.* Top: illustration of collagen being secreted from a fibroblast. Ribosomes (represented by small dots) synthesize and deposit collagen into the endoplasmic reticulum (ER). The collagen travels to the Golgi and then to vesicles, which fuse with the plasma membrane to release collagen outside the cell. The triple helix structure of collagen then assembles into large bundles, as seen using electron microscopy (Bottom, from https://www.sciencephoto.com/media/996019/view/collagen-em)

fibroblast in the right direction. When they arrive at the wound, they use their ability to grip ECM in the environment to contract the edges of the wound as other processes also go to work (e.g., blood clotting). Upon closing the wound, fibroblasts will return to secreting molecules to fill in gaps. Not only will they make new ECM, but now they will also secrete other proteins that serve to break down the clots that sealed the wound. The collagen (and other types of ECM) that fibroblasts make during wound healing is organized differently than under typical conditions. It tends to not spread out as much or be as elastic, leading to what we call scar tissue. Fibroblasts are at the center of tissue remodeling approaches for the medical treatment of burns and ulcers. These types of injuries have killed the local fibroblasts, so healthy ones are collected from unaffected tissues and injected at the wounded site to manage reconstruction.

7

Neurons

In order for multicellular life to succeed, there must be communication between cells. This is easiest for near neighbors; one cell can send a chemical message and count on random diffusion to reach cells in the local vicinity. Messages that need to be sent much further can instead enter the bloodstream and travel throughout the body within a minute (a specialty of endocrine cells, discussed in Chap. 9). While that is pretty fast, in many cases, we need cells to communicate information over longer distances immediately. Neurons are exceptionally good at accomplishing this demand by transmitting information via electrical currents over very long projections extending from their plasma membranes. These currents can travel over 100 meters/sec, which is important if you have just touched a hot stove or seen an oncoming car about to cross your path.

Neurons exist in a few different configurations, but all contain a typical cell body that is home to the nucleus and other organelles. Most neurons additionally contain a unique cellular appendage called an *axon*, which can extend long distances toward cells that the neuron communicates with (Fig. 7.1). For example, the sciatic nerve, which is a bundle of axons from thousands of neurons, covers the distance from your lower back to the top of your foot. The extension of an axon uses microfilaments to push the plasma membrane forward, just like we saw with crawling fibroblasts in the previous chapter, except that in these neurons, the cell body stays put while just a discrete region (called the growth cone) is pushed forward (a video of an active growth cone is found here: https://www.youtube.com/watch?v=op_LdPDVgNo).

Once established, the axon sends electrical currents away from the cell body of the neuron. The end, or terminal, of the axon forms a connection with one or more cells that it communicates with. Depending on the type of

L. Saucedo, *Getting to Know Your Cells*, https://doi.org/10.1007/978-3-031-30146-9_7

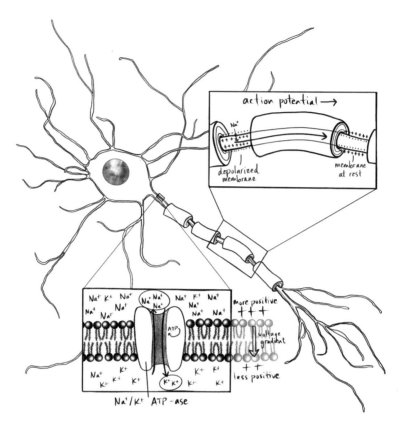

Fig. 7.1 *Nerve transmission.* A neuron with numerous dendrites and one myelinated axon extending from the cell body. A voltage difference is established by Na⁺/K⁺ ATPase pumps, such that in a resting state, the interior of a cell is less positive than outside the cell. A change in this voltage difference (called depolarization) triggers an action potential to travel along the axon, creating a current. The speed of the current is accelerated when the axon is myelinated because this insulation exempts large sections of the membrane from depolarization

neuron, the recipient cell can be another neuron, an endocrine cell, or a muscle. Upon reaching the end of the axon, the current leads to the release of molecules, called *neurotransmitters*, into a small gap of space (~20 nanometers) between it and the recipient cell, called a *synaptic cleft*. The neurotransmitters quickly bind to receptors on the recipient cell to evoke a response.

Currents are simply movements of charged particles. In cells, charged atoms called *ions* serve as the elements of electric flow. Some of these ions are familiar components of your diet: sodium (Na^+), potassium (K^+), calcium (Ca^{2+}), and chloride (Cl^-). Cells bring in fewer of the positive (+) ions than available, resulting in the cell being comparatively negative (−) to the extracellular environment and creating a voltage difference across the plasma membrane (Fig. 7.1). The hydrophobic chemistry within membranes that serves as a barrier to water is even less friendly to charged particles, and therefore protein transporters are essential to providing passage to ions. Additionally, the unequal distribution of any molecules, including ions, requires energy. In most cases, this energy is provided by adenosine triphosphate (ATP) (as discussed in Chap. 5). A direct connection between ATP use and ion movement is nicely illustrated by a protein transporter found on the plasma membrane of all animal cells, called the Na^+/K^+ ATPase pump. Over 1/4th of the ATP that the body produces is used to fuel this ubiquitous pump, which pushes 3 Na^+ ions out of the cell for every 2 K^+ it pulls into the cell (Fig. 7.1). This uneven exchange of positively charged ions greatly contributes to the interior of cells being relatively negative and creates the groundwork for conducting electrical currents.

Neurons are in a *resting state* until they are needed. In terms of voltage, this is measured as -70 millivolts. A neuron sits in wait until a signal is received. These signals can be a chemical, often a neurotransmitter, but can also be a physical force (e.g., a full bladder) or a change in temperature. Typically, these signals are received at the plasma membrane around the cell body of the neuron. Sometimes, that membrane extends out into several small projections called dendrites (Fig. 7.1), increasing the distance and number of points for the neuron to receive communication. Whatever the stimulus is, the receptor that detects it will change its shape and briefly opens up a channel that allows ions to move across the membrane. This type of protein is aptly named a *gated ion channel* (Fig. 7.1). If the signal is meant to activate the neuron, the channel will allow Na^+ to cross the membrane. Recall that a lot of energy was previously dedicated to pushing Na^+ out of the cell by the Na^+/K^+ ATPase pump. Given the opportunity, Na^+ ions will reenter until they are evenly distributed across the membrane. As they do this, the voltage inside the cell near that particular receptor will increase. If it rises from -70 mV to -55 mV, this will trigger the neuron to "fire," that is, to commit to propagating a series of voltage changes from the initial point all the way to the end of the axon. Upon reaching -55 mV, another ion-gated channel will open that temporarily allows even more Na^+ to cross the membrane. This next channel does not bind any chemicals but changes shape in response to local changes in voltage. Now, so much Na^+ enters that the local voltage within the cell reaches +30 mV or higher. When the voltage within the cell relative to the outside of the cell switches from being negative to positive, it is called *depolarization*. It is immediately followed by *repolarization*, thanks to one more ion-gated channel. This channel changes shape in response to the dramatic voltage change and opens to allow K^+ to cross the membrane. Similar to Na^+, now K^+ can attempt to undo the previous work of the Na⁺/K⁺ ATPase pump. And so it rushes out of the cell, bringing the local voltage to even lower than -70 mV. These sudden changes in voltage are collectively called *action potential*. One important thing to recognize is that while this region of the membrane has repolarized, Na⁺ and K⁺ ions are on the wrong side of the membrane compared to the resting state. This means that if a new signal is received, it cannot trigger another action potential. This region of the membrane is now in a *refractory period* and will remain insensitive until the Na⁺/K⁺ ATPase pump reestablishes the unequal distribution of Na⁺ and K⁺ ions. This recovery period is helpful because it helps ensure that the current moves further along the membrane until it reaches the end of the axon (Fig. 7.1). Notably, as the additional gated ion channels open along the membrane, no energy is required – all the ions are moving down the *gradient* previously created. This is a great means to boost speed.

For longer axons, another means to accelerate transmission comes into play. Rather than needing to redo every step of an action potential down the entire length of the axon, long stretches of the membrane can be "skipped" over. This is possible when parts of the axon become *myelinated*. Myelin contains a lot of fat, which prevents the ion differences across the membrane from being sensed – basically serving as electrical insulation. Only the small regions of the axon that remain unmyelinated need to undergo action potentials, with the current traveling between them for most of the length of the axon. Loss of myelination is a characteristic of several disorders, sometimes temporarily seen with most cases of Guillain-Barré syndrome (GBS) and sometimes permanent as with multiple sclerosis. As expected, the transmission speed of nerve signaling slows down, and this leads to a wide range of symptoms in patients.

Upon reaching the axon terminal, a new gated ion channel comes into play. In response to the local depolarization, this channel allows Ca²⁺ to enter the cell. In turn, the influx of Ca²⁺ will cause neurotransmitters to leave the neuron.

Neurotransmitters are synthesized in the secretory pathway (as we saw in the preceding chapter), located in the cell body of the neuron, and then delivered within small vesicles to the axon terminal via cytoskeletal highways and *motor proteins*. A motor protein consists of a pair of intertwining halves, each having a "head" and a "tail." The tail is what holds onto the cargo as the heads take turns binding and releasing from the cytoskeleton, analogous to taking steps (Fig. 7.2). The component of the cytoskeleton that serves as the highway along the length of the axon is called a *microtubule*. In spite of the "micro" prefix, it is the largest element of the cytoskeleton, which has a hollow center with 13 strands of proteins aligning to serve as an outer wall (Fig. 7.2). The

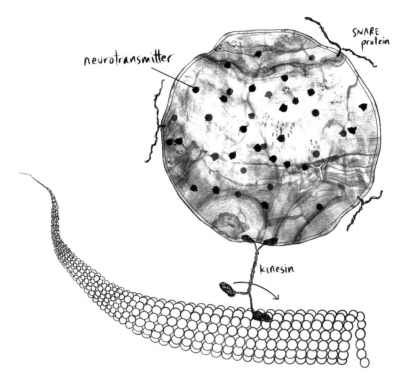

Fig. 7.2 *Motor protein carrying cargo on a microtubule*. Microtubules serve as the highway for transporting vesicles containing neurotransmitters from the cell body of a neuron to the end of its axon. The motor protein, kinesin, has the ability to take steps along a microtubule while bound to vesicles. When Ca^{2+} is released into the cell, this activates SNARE proteins on the vesicles, which will initiate fusion with the target membrane to release the neurotransmitters into the synapse

class of motor proteins that carry neurotransmitters toward the end of the axon is called *kinesins*. Delivery to the proper location is not all that is needed, though; the neurotransmitters are like an unopened package on your porch. This is because the membranes of the vesicles packed with neurotransmitters contain proteins (called soluble N-ethylmaleimide-sensitive factor attachment protein receptors (SNAREs)) that need to latch on and promote fusion to the plasma membrane. But these proteins are in a dormant state until they are bound to Ca^{2+}. Normally, Ca^{2+} is kept out of the cell (or retained within intracellular organelles), thanks to Ca^{2+} ATPase pumps, and so neurotransmitters are stored until a depolarization event allows an influx of Ca^{2+} and triggers their release. The neurotransmitters then enter the synaptic cleft and can bind to the next cell. Therefore, nerve transmission often starts and ends with a neurotransmitter, a molecule that confers specificity but is slow to diffuse. However, the vast majority of the distance that the information must travel is via electrical energy spanning the length of the axon.

Several dozens of different neurotransmitters have been identified in humans. One way to group them is determining whether they are excitatory or inhibitory. This reflects whether the binding of the neurotransmitter on a target cell elicits a depolarization event (excitatory), as described in this chapter, or causes the target cell to become more resistant to depolarization (inhibitory). Inhibitory neurotransmitters are able to desensitize a cell because binding to their receptor opens up an ion channel that lets Cl^- into the cell. This reduces the relative voltage inside the cell so that its resting state drops even lower than -70 mV. Thus, the next excitatory neurotransmitter to bind is less likely to cause the cell to reach the threshold of -55 mV needed to evoke a response. Unfortunately, just knowing the name of a neurotransmitter does not necessarily reveal if it is excitatory or inhibitory. Instead, it will be dependent on the receptor expressed by the cell that binds the neurotransmitter. For example, acetylcholine acts as an excitatory neurotransmitter on numerous cell types, including skeletal muscle (leading to contraction, as described in the next chapter), but as an inhibitory neurotransmitter for cardiac muscle, reducing heart rate.

The function of neurotransmitters is frequently targeted by pharmaceuticals. They can work by many different mechanisms, including increasing or decreasing the amount of signaling by neurotransmitters. For example, some antianxiety medicines increase the amount of time a neurotransmitter called serotonin is able to remain in the synapse and thus bind receptors to stimulate target neurons. In contrast, anesthetics like lidocaine prevent neurons from opening their voltage-gated Na^+ channels, thus preventing nerve transmission, which would relay information about pain.

Neurons convey information from the environment, process emotions, form memories, and direct physical movement. Disruptions in nerve signaling thus manifest in many different and profound ways. Multiple neurotransmitters are implicated as being altered in Alzheimer's disease and Parkinson's disease. In amyotrophic lateral sclerosis (ALS), motor neurons degenerate, weakening muscles until they become paralyzed. While extremely rare, sensory neurons of people with congenital insensitivity to pain (CIP) fail to function. Navigating the world without pain to alert you of injury is dangerous, and the lifespan of these patients is often dramatically shortened.

While most of the cell types discussed in this book are essential to our survival, neurons may be the ones we primarily define ourselves by when we think of who we are. Uploading the information stored in our brains, which at the cellular level is a distinct pattern of action potentials, is a common theme in science fiction as a means of preserving our existence beyond life.

8

Muscle

Directed movement of individual cells or the components within them is not unusual, but the ability of cells to move the organism they occupy is unique to muscles. Muscles can be divided into skeletal, cardiac, and smooth. Cardiac and smooth muscles share the primary function of squeezing. While cardiac muscle specializes in squeezing your heart, smooth muscle squeezes a wide range of organs, such as your blood vessels, intestines, bladder, and lungs. In contrast, skeletal muscle pulls/pushes bones. While the output differs, the same intracellular mechanism of contraction is at play and mediated by the cytoskeleton. Muscle mass is frequently associated with strength, and indeed the expression "what doesn't kill you makes you stronger" is biologically relevant to this cell type.

The process of contraction is easiest to visualize in skeletal muscle. This is because this muscle type is the result of hundreds of individual cells, called *myoblasts*, fusing together to form large, long fibers with patterned arrays that are noticeably different when contracted or relaxed. Following rapid proliferation during embryogenesis, myoblasts organize themselves into rows, and then the plasma membranes between individual cells merge (Fig. 8.1). This creates a *myofiber*, containing multiple nuclei from the contributing cells. Additional myoblasts fuse as the developing muscle grows. As the muscle fiber differentiates, the bulk of the cytoplasm fills with intracellular fibers called *myofibrils*. Some of these intracellular fibers consist of microfilaments, the component of the cytoskeleton that pushes on the plasma membrane as a cell crawls or extends cellular appendages, as we saw in Chaps. 6 and 7. However, in contrast to their contribution to crawling, the microfilaments found in muscle are not growing but instead are of a fixed length.

© The Author(s), under exclusive license to Springer Nature Switzerland AG 2023
L. Saucedo, *Getting to Know Your Cells*, https://doi.org/10.1007/978-3-031-30146-9_8

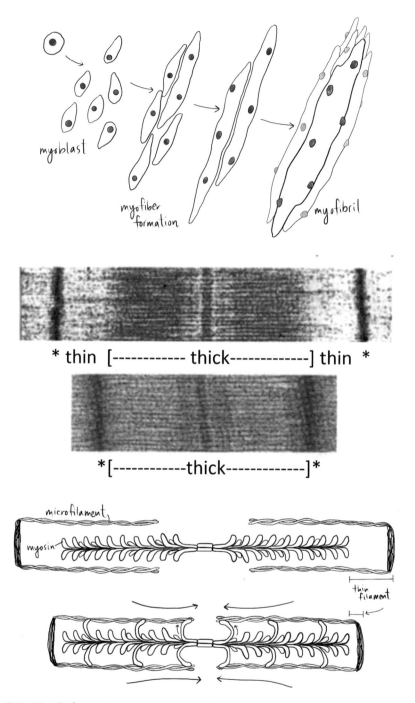

myoblast

myofiber
formation

myofibril

* thin [------------ thick-------------] thin *

[------------thick-------------]

microfilament

myosin

thin
filament

Fig. 8.1 *Muscle formation and contraction.* Skeletal muscle forms through the fusion of myoblasts and the subsequent alignment of myofibers (top illustration). The banding pattern of relaxed and contracted skeletal muscle can be seen with microscopy (middle, * indicates the edges of the contractile unit, adapted from https://aworldofbiology.weebly.com/muscle-and-movement.html). This contraction is due to chains of immobile motor proteins called myosins, which are binding and sliding microfilaments toward each other (bottom illustration)

Dozens of strings of microfilaments run lengthwise in muscle fibers and are described as "thin filaments" due to their relative width as compared to a second abundant structure called "thick filaments." Thick filaments consist of the motor protein *myosin*. Again, in contrast to the role of motor proteins traveling along the cytoskeleton, as described in neurons (Chap. 7), these myosins are stuck in place. The tails of individual myosin proteins associate to form long chains. And then two chains of myosin attach in opposing directions. This sets the stage for the heads of the myosin to interact with neighboring microfilaments. But now, the movement of the heads can no longer propel the myosin forward, and so the kinetic energy results in pulling the microfilaments together instead, leading to contraction and making it appear as if the thin filaments have disappeared (Fig. 8.1). Because skeletal muscle develops by the fusion of individual cells, this contraction happens over a long distance, and the change in the banding pattern of thin and thick filaments can be seen using an electron microscope (Fig. 8.1).

Cardiac and smooth muscles use microfilaments and myosin to contract as well, but a stark difference is that the cells involved remain individualized. This means that the contraction of one cell needs to be coordinated with the contractions of its neighboring cells. This is accomplished by a structure called a *gap junction*, which enables cells to share components of their cytoplasm via a tunnel between adjacent plasma membranes. This structure will be illustrated in the next chapter, but it reveals why the contractions of these two muscle types spread like waves – as the stimulus diffuses from one cell to its neighbors via gap junctions, contractions happen in succession. Remarkably, these waves of contraction can be seen in stem cells that have been coaxed to differentiate into cardiac muscles using tissue culture technics as seen here: https://www.youtube.com/watch?v=IlDoslXHdG0.

In addition to resulting from the fusion of individual cells, skeletal muscle stands apart from cardiac and smooth muscles because you get to decide when they contract, meriting them the term "voluntary muscle." In contrast, cardiac and smooth muscles contract without your permission and are thus "involuntary." In either case, they are initiated by neuron signaling. Neurons signal to muscles in different configurations – for example, skeletal muscles have distinct neuromuscular junctions, while smooth muscles are overlaid by a diffuse blanket of neurons. For many involuntary muscles, the neurons fire in a self-sustaining rhythm, earning them the description as "pacemaker neurons." Similar to neuron-to-neuron signaling (Chap. 7), when a neuron signals to a muscle, it causes a redistribution of ions. The key ion that leads to muscle contraction within a muscle is calcium (Ca^{2+}). Muscle cells contain a specialized version of the endoplasmic reticulum, called the *sarcoplasmic reticulum*, which stores Ca^{2+} within its membranes. This keeps levels of Ca^{2+} low in the cytoplasm, where the contraction machinery (microfilaments and myosins) is located. In the relaxed state, the myosins' heads are unable to interact with microfilaments because a molecule called *tropomyosin* is coating the microfilaments. Upon the release of Ca^{2+} into the cytoplasm, the Ca^{2+} binds to tropomyosin, shifting its position so that myosin can now bind and pull the microfilaments together (Fig. 8.1). Sometimes our voluntary muscles will contract in an involuntary manner – we perceive this as an annoying twitch. Numerous causes for twitching have been proposed, but anything that can disrupt ion gradients in either neurons or muscles can result in twitching.

Muscles are the one type of tissue that we often purposely injure; this is how we gain muscle mass as adults. Exercising muscles can cause damage to myofibrils; when they are repaired, they are often rebuilt to be thicker. In addition, *satellite cells*, which reside near muscle fibers, can be stimulated to divide and then fuse with existing myofibrils. While this repair process can be associated with soreness, in contradiction to the adage "no pain, no gain," research indicates that it is not necessary to exercise that intensely to stimulate muscle growth. While exercise certainly affects muscle size, genetics has a role too. Over 40 genes have been implicated to contribute to variation in muscle growth in mammals. *Myostatin* is perhaps the most dramatic of those genes, with a central role in inhibiting muscle growth by countering myoblast proliferation and fusion. Mutations that inactivate myostatin result in excessive muscle, as demonstrated by naturally occurring "double-muscled" cattle breeds and in several recently genetically engineered animals of agricultural interest. Some humans with relatively excessive muscle mass have been identified as similarly having genetic differences that prevent myostatin from fully functioning. Not surprisingly, uncovering ways of increasing muscle mass is of interest to a wide variety of people but could be especially important in improving the quality of life for elderly people. Muscle mass starts to decrease in one's 30s, even in the physically active. By age 70, it is not unusual to have lost 25% of muscle, putting people at higher risk for falls and fractures. Multiple clinical trials using pharmaceuticals to counter myostatin function have been performed. While small gains in muscle mass have been reported, no significant increases in strength have yet to be achieved. This finding mirrors research done using mice genetically engineered to lack myostatin; increased muscle size in these animals actually resulted in less strength. It appears that the extra muscle that develops is compromised in function.

On the other end of the spectrum, extreme muscle-wasting diseases can be inherited and are called *muscular dystrophies*. Duchenne muscular dystrophy (DMD) is the most common and is the result of a mutation in the largest gene found in humans, which codes for a production of a protein called dystrophin. The role of dystrophin is to secure the myofibrils to anchors in the plasma membrane, which in turn are bound to large structural proteins outside the muscle fibers. Similar to the intermediate filaments that we saw contributing to the function of epithelial cells in Chap. 4, dystrophin serves to protect against mechanical stress. While intermediate filaments absorb stress emanating from outside the cell, dystrophin counteracts damage from the force of contraction within the muscle fiber. When dystrophin function is compromised, as it is in DMD, the myofibrils break down.

It is probably pretty obvious that muscle movement also requires a lot of energy. During contraction, adenosine triphosphate (ATP) that is bound to the heads of myosins plays a key role. The energy that is stored in the chemical bonds of ATP is released (by breaking bonds) and transformed into kinetic energy to power the movement of myosin pulling on neighboring microfilaments. Given this essential role, it might seem surprising to learn that in a muscle type called "fast twitch," a very inefficient means of producing ATP is used. We previously learned how aerobic respiration can convert a single sugar into 30 ATPs via a series of steps that are dependent on both oxygen and mitochondria (Chap. 5). However, cells can also perform anaerobic respiration or *fermentation*, a process that produces only two ATPs from a starting sugar. Most of the energy remains in the products of fermentation, which can be either lactic acid or alcohol, an outcome that is exploited during the production of many foods that we consume, like yogurt and wine. When human cells undergo fermentation, we only produce lactic acid, which keeps us from getting drunk from a workout. Fast-twitch muscles differentiate to favor fermentation over aerobic respiration by containing few mitochondria within them. In addition, they are surrounded by minimal capillaries, which limit local oxygen levels. Fast-twitch muscles rapidly deplete sugar supplies since so little ATP is produced, and thus they fatigue quickly. Therefore, they are primarily used for short bursts of activity, like jumping or sprinting. For sustained activity, you will need your slow-twitch muscle fibers, which are optimized to perform aerobic respiration. However, if oxygen levels become insufficient, slow-twitch muscles can shift to fermentation as well. When ATP levels become too low, usually because available sugar has become scarce, muscles get stuck in a contracted position. This is because myosins cannot "let go" of microfilaments (Fig. 8.1) until a new ATP binds. During exercise, this can manifest as a cramp, but it is also the basis for rigor mortis following death, when ATP is permanently depleted. Once again, the phrase "what doesn't kill you, makes you stronger" applies quite well to muscle cell biology.

9

Endocrine Cells

It is common to hear people blame their hormones for a multitude of things, from behavior to weight gain/loss. Indeed, hormones play key roles in modulating emotions and metabolism. They are produced by cells located in one of a dozen endocrine glands and are able to enter the bloodstream, which allows them to serve as a messenger to any other cell in the body. However, only cells that express the receptors to bind a hormone are able to "hear" or respond to that hormone. Additionally, the release of hormones is frequently in response to other inputs that we often have some control over, ranging from the amount of light in our environment to the food we consume. In some cases, focusing only on hormones is truly blaming the messenger.

As mentioned in Chap. 4, endocrine cells are a specialized type of epithelial cell. During early development, they move from the surface of an embryo inward, forming small burrows. Some will maintain a route to the environment via a tube remaining after invagination, such as the endocrine cells of sweat and salivary glands. But many will have that passageway sealed off, and thus their secretory products remain within us. Cells of an endocrine gland have the ability to receive a signal that evokes the release of hormones. Until then, the hormones are packed into secretory vesicles sitting in wait. Often, that signal comes from another endocrine gland. As you will soon see, the response of endocrine cells to their signals is complex but shares biological mechanisms with neurons and muscles that we have already encountered in previous chapters. In this chapter, we will focus on the specific endocrine cells found in the pancreas and ovaries and look more closely at how these cells use the hormones insulin and estrogen, respectively, to evoke cellular responses.

L. Saucedo, *Getting to Know Your Cells*, https://doi.org/10.1007/978-3-031-30146-9_9

The pancreas is located behind the stomach and has a small tube connecting it to the beginning of the intestine. As briefly noted in Chap. 5, it is able to modulate blood glucose levels by secreting insulin and glucagon. Here we will look more closely at the role of insulin. First, the pancreas needs to sense glucose levels. It does this by having a glucose transporter on the plasma membranes of specific subsets of cells in the pancreas, called beta cells (Fig. 9.1). If there is relatively more glucose in the bloodstream than in the cell, glucose will move into the cell via this transporter. Glucose will then be processed via aerobic respiration (covered in Chap. 5), leading to a rise in adenosine triphosphate (ATP) levels. High levels of ATP will next bind to and inhibit the function of another transporter on the plasma membrane, which normally allows potassium (K^+) into the cells. Preventing the influx of K^+ alters the relative voltage of the cells and trips a voltage-sensing Ca^{2+} gate. This allows Ca^{2+} to flood into the cell, and just like we saw with neurotransmitters, high levels of Ca^{2+} in the cytoplasm will enable the secretory vesicles containing insulin to fuse to the plasma membrane.

Importantly, the secretion of insulin from one beta cell can be coordinated with neighboring cells, thanks to the gap junctions between them (Fig. 9.1). Similar to how cardiac and smooth muscle cells share components of their cytoplasm to cause waves of contraction (Chap. 8), gap junctions allow the Ca^{2+} to pass between beta cells. These junctions are formed by a protein aptly called *connexin*, which spans the plasma membrane. Six copies of connexin arrange to form the shape of a hoop, in contact with the plasma membrane on the outer edge but leaving a small pore in the middle (Fig. 9.1). The extracellular surfaces can then connect between cells, and the aligned pores serve as a tunnel to connect cytoplasm. Now, beta cells, in addition to the initial one that received a signal, can respond, amplifying and expediting the response. Insulin is released into the bloodstream.

Pancreatic beta cells take in some glucose in order to know when to release insulin, but it is our liver and muscle cells that are specialized to remove excess amounts of glucose from our bloodstream and serve as storage sites. Thus, they are the key cells to respond to insulin. Like the majority of signaling molecules meant to evoke a cellular response, insulin does not enter its target cells. It can only bind to the extracellular portion of its receptor that is embedded in the plasma membrane of responsive cells. That means the receptor must somehow relay the information of the presence of insulin to the interior of the cell. The ability to accomplish this is known as *signal transduction*. A common theme of protein receptors is that they not only span the width of the plasma membrane but also have substantial portions facing the extracellular and intracellular environment. The extracellular portion will be shaped

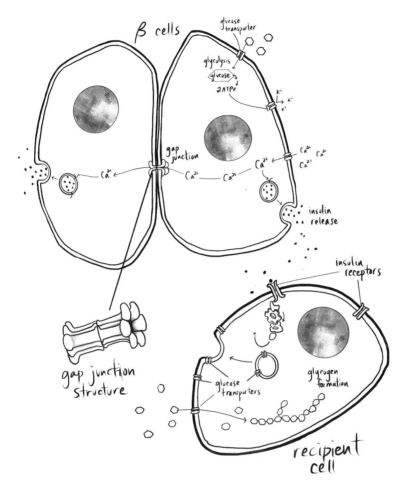

Fig. 9.1 *Signaling between beta (β) cells of the pancreas and a recipient cell.* When blood glucose levels are high, pancreatic beta cells respond by releasing insulin. This involves a series of events: importing and metabolizing glucose into ATP, which then suppresses K^+ import. This alters the cellular voltage, which subsequently triggers Ca^{2+} import. Heightened levels of intercellular Ca^{2+} then promote the fusion of secretory vesicles containing insulin to the plasma membrane. This response is made more robust by gap junctions that allow Ca^{2+} to be shared between neighboring beta cells, thus skipping over all prior steps. The insulin released by multiple cells can then travel throughout the body to prompt cells with insulin receptors to import much more glucose for storage as glycogen. This again involves a cascade of signals within the cell that leads to the fusion of vesicles (with glucose transporters) to the plasma membrane

to specifically bind the signal, such as insulin, outside of the cell. But this binding interaction causes a change in the shape of the receptor, including the portion of the receptor that faces the inside of the cell, thus transducing the information to the interior of the cell (Fig. 9.1). In the case of the insulin receptor, this change in the shape of the intracellular portion allows proteins that are normally freely diffusing through the cytoplasm to instead attach to the insulin receptor. This concentrates a number of proteins into a discrete region near the plasma membrane. Once in close proximity to one another, some are able to chemically modify one another. These modifications will lead to a series of changes that allow intracellular glucose transporters to fuse to the plasma membrane. Now, glucose will move out of the bloodstream and into hepatocytes and muscles. However, insulin signaling does not stop there. This glucose is meant to be stored, and so downstream of the insulin receptor, additional chemical modifications of proteins will activate enzymes to connect individual glucose molecules together to form the complex carbohydrate glycogen. While it is probably evident why we do not want blood glucose levels to drop too low – our neurons and muscles are especially impacted if they cannot use glucose to produce the vast amount of ATP they need – why is it a problem to have high levels of glucose in our blood? A serious consequence is that the increased concentration of glucose damages the blood vessels over time. This affects blood flow, which can then manifest as cardiovascular disease, organ damage (especially the eyes and kidneys), and necrosis of tissues (especially those located farthest away from the heart toes and fingers).

Ovaries are both a reproductive organ and an endocrine gland. Specifically, granulosa cells inside the ovary, which surround the oocyte, are the cell type responsible for producing the most estrogen. Estrogen is derived from cholesterol via a couple of chemical modifications, including testosterone, as an earlier product in the synthesis pathway. Cholesterol, estrogen, and testosterone are all steroids – small, flat molecules that are mostly hydrophobic, which allows them to pass easily across the phospholipid tails of membranes (Chap. 2). Thus, estrogen cannot be stored within a membranous vesicle prior to release, like insulin. Instead, it is synthesized and immediately released when needed. Estrogen is an example of a hormone that is produced in response to other hormones. Two key hormones in the control of estrogen levels are follicle-stimulating hormone (FSH) and luteinizing hormone (LH), both produced by the pituitary gland. FSH and LH increase estrogen by accelerating the production of enzymes and reactants needed for its synthesis. Estrogen then moves down its concentration gradient and leaves granulosa cells for the bloodstream.

Numerous cells are able to respond to estrogen, including cells found in the reproductive tract, cardiovascular system, brain, breast, and bone. Much of the understanding of the numerous roles of estrogen has been elucidated because of studies on menopause – when estrogen production in the ovaries becomes diminished. While small amounts are still produced by the adrenal gland and other cells, like adipocytes and neurons, it is clear that many characteristics of menopause can be countered by estrogen supplements (also known as hormone replacement therapy (HRT)). The two targets of estrogen signaling that we will look at closely are the breast and bone.

Because estrogen can cross the plasma membrane, receptors for estrogen can be found in the cytoplasm. These receptors are called "nuclear receptors" because after binding to estrogen, they are able to enter the nucleus, interact with deoxyribonucleic acid (DNA), and alter gene expression. Interestingly, even though it is not a barrier for estrogen, some receptors for estrogen are found embedded in the plasma membrane. In this case, estrogen does not enter the cell but instead initiates signal transduction pathways, which ultimately affect the transcription of genes as well. Target genes can stimulate proliferation, migration, differentiation, or *apoptosis* (an active process of cell death illustrated in the next chapter), depending on the specific cell type and which other genes are being expressed. Estrogen signaling is often exploited by breast cancer cells, with receptors that promote excessive cell division and migration. Referred to as ER+ (for *e*strogen *r*eceptor positive), this type of breast cancer can be impeded by treatment with molecules that mimic estrogen but do not elicit a cellular response. One such estrogen mimic is called tamoxifen. First synthesized in trials to create oral contraceptives in the 1960s, it was tossed aside when studies showed it stimulated rather than inhibited ovulation. However, over a decade later, it was approved as a treatment for metastatic breast cancer. After being metabolized by p450s in the liver (as discussed in Chap. 5), tamoxifen competes with estrogen to bind estrogen receptors. But when it binds, tamoxifen does not cause the same change in the shape of the receptor needed for it to induce the expression of genes that promote proliferation in breast epithelial cells. This response is cell-type specific, as tamoxifen does promote the proliferation of endometrial cells, and so long-term treatment increases the chance of endometrial cancer. This underscores a key feature of cell communication that we have seen before with neurotransmitters (Chap. 7); a molecule serving as a messenger can "tell" different cells to do very different things.

Menopause has long been correlated to bone deterioration, and thus estrogen's role in bone integrity has been carefully examined. Studies indicate that in both adult women and men, estrogen is the key hormone in maintaining bone health. Bones may seem like static structures, but they are living organs that are continuously removed and replaced, albeit at a relatively slow rate compared to other organs – it is estimated that most of the adult skeleton is replenished about once a decade. Recall that bone is a type of connective tissue inhabited by fibroblasts (Chap. 6). Two types of specialized fibroblasts in bone are called osteoclasts, which do the job of breaking down existing bone to clear space for new bone and osteoblasts, which do the work of building up that new bone. Both are found along the surface of bones and contain estrogen receptors. Estrogen inhibits apoptosis and extends the lifespan of osteoblasts. In addition, estrogen blocks the function of osteoclasts by preventing differentiation and inducing apoptosis. Thus, reduced estrogen leads to fewer osteoblasts to generate new bone while increasing the number and activity of osteoclasts, which break it down. Estrogen's fundamental task in preserving bone strength is also supported by the reduction in bone density in young women taking oral contraceptives without estrogen and in men who have a genetic mutation that prevents them from producing estrogen. In all cases, reintroducing estrogen minimizes bone loss.

Endocrine cells are similar to neurons in their ability to send a signal over a long distance. However, hormones take their time, relatively, traveling along the circulatory system. On the other hand, hormones have much more flexibility with regard to which target cell they encounter, unlike nerve signaling, which involves a direct bridge between two cells. Over evolutionary time, it seems likely that the "choice" between using endocrine cells and neurons for sending a long-distance message is based on whether the message must be delivered immediately and precisely (I am touching a hot stove ➜ neurons) or not (time to store away sugar ➜ endocrine cells).

10

White Blood Cells

"Know thyself" is a sentiment that has been an aspiration among humans for at least a millennium. But our immune system is already exceptionally adept at this skill. Since it is impossible to predict every infectious agent that you may encounter (many of which are often rapidly evolving), an efficient way to shape an agile defense is to dedicate efforts to recognize what is "self" and then target and destroy what does not match as the default response. There are many arms to the immune system, and we will see how some of them work in concert with other cell types in Chap. 13, but this chapter will focus on the abilities of certain key white blood cells (WBCs).

To start, WBCs are not actually white. However, when blood is separated by density using a process called *centrifugation*, white blood cells form a clear layer between red blood cells (RBCs) and yellow plasma (the liquid component of blood). WBCs normally account for only about 1% of the blood but have a much more active role than red blood cells. Therefore, unlike their red counterparts, WBCs keep their nuclei as they differentiate, allowing them to be dynamic and responsive to their environment, thanks to expressing different genes as needed. There are several types of white blood cells, but this chapter will focus on two subtypes that are the most individualized for each person: B cells and T cells. These types of WBCs are not satisfied with only detecting the same invaders that threatened our ancestors but are instead tailored to "look" for novel intruders that you encounter in your lifetime.

Because they tend to huddle up in lymph nodes when not circulating in our blood, B and T cells are part of the category of WBCs called lymphocytes. They are derived from the same stem cells in the bone marrow as RBCs (Chap. 2), and that is the location where B cells will complete their differentiation. In contrast, T cells will travel to the thymus for a final step of development. B and T cells use receptors on their plasma membranes to detect unfamiliar material, simply called B-cell receptor (BCR) and T-cell receptor (TCR). But these receptors are remarkable because the genes that encode them are deliberately broken and reformed by cellular machinery in order to rearrange the DNA sequence in novel ways. This allows for each cell to randomly encode a different receptor. It has been estimated that a person has at least 100,000,000 different receptors on their B and T cells. That suggests that these receptors have the capacity to bind to an equal number of unique targets as they circulate. Importantly, the arbitrary construction of receptors means that they could bind to our own cells and proteins; thus, we need a way to protect against an immune response to "self," known as *autoimmunity*. A well-understood process to safeguard from this inappropriate outcome takes place in the thymus, an organ located between the top of your lungs. Within the thymus, T cells undergo a strict selection process. Thymic cells use receptors on their plasma membranes to display a wide range of protein fragments that can be produced by each person's genome. If a passing T cell is able to bind strongly to any, that will elicit a response in the T cell to initiate its own death. Known as *apoptosis*, and briefly noted in the previous chapter, most cells in our body are poised to self-destruct in a highly controlled manner (Fig. 10.1; a video of the process can be found here: https://www.youtube.com/watch?v=rs1Je-8Y3Po).

Enzymes, called *caspases*, that degrade essential cellular proteins lay in wait for a signal that the cell is not healthy or even dangerous to the organism. As the caspases break down a cell, the remnants remain enclosed by membranes, but within small enough packages to be digested by other cells. A signal to activate caspase activity can arise from within a cell (for example, damaged DNA) or from outside a cell, such as the tight binding between receptors on a thymic cell and a T cell. Over 95% of T cells that enter the thymus are eliminated through this process. Consequently, T cells that would be able to recognize anything that is self should not be allowed to persist. Selection against B cells that recognize self occurs as well, though it is ongoing throughout multiple stages of development, occurs in multiple locations, and results in more varied outcomes: for example, instead of eliciting apoptosis, a B cell can instead edit the BCR it expresses until it can no longer bind any resident molecules.

Infectious material can either be "free" or "hidden" within our cells. For example, some bacteria always remain free while they reproduce within our bodies, but other bacteria and all viruses need to enter our cells in order to reproduce. B and T cells focus their attention separately on these two

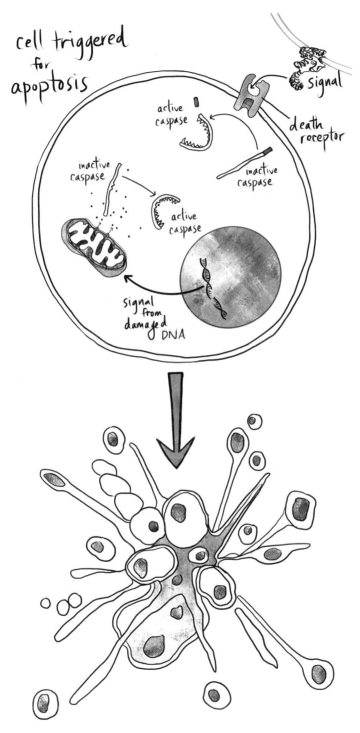

Fig. 10.1 *Apoptosis*. Cells can receive signals from neighbor cells or from internal damage that activate caspase enzymes to break the cell down into small, membrane-bound globules that are subsequently digested by other cells

alternatives. The surveillance of infected cells, also known as "host cells" (for the infectious agent) is a task for T cells. Almost all of our cells are able to display small fragments of proteins within them, similar to thymic cells during T-cell selection. However, a different receptor is used with the express purpose of revealing if the cell is infected. This receptor is called major histocompatibility complex type 1 (MHCI). As MHCI moves through the secretory pathway (described in Chap. 6), it is loaded with an intracellular protein fragment. Upon embedding into the plasma membrane, this fragment now faces the outside of the cell. TCRs on T cells are able to briefly dock with MHCIs on all other cell types. If the protein fragment belongs to us, the binding between the two receptors will be weak, and nothing happens. But if the protein fragment is not local and fits the TCR receptor, the binding will be strong and elicit apoptosis of the infected cell (Fig. 10.1). In short, the tables have turned for T cells that are allowed out of the thymus. Instead of committing suicide when their TCR finds a binding partner, T cells now stimulate this outcome in other cells, earning them the term "cytotoxic" T cells.

Another set of T cells will instead take on the role of "helper" T cells. Helper T cells aid in many steps of the immune response, including assisting cytotoxic T cells, but they also play a major role in promoting B-cell function. If a B cell encounters something unfamiliar that binds well to its BCR, a T helper cell that recognizes the same material stimulates the B cell to divide profusely. This collaboration is much more likely to happen if the matching B and T cells are not being pushed through the bloodstream, and this is why it is important for these cells to frequently enter and rest in lymph tissues where more stable interactions can take place. Importantly, most of the resulting progeny B cells will convert their BCR to antibodies. Instead of remaining embedded in the plasma membrane, these antibodies, which can bind the same material as the parent cell's BCR, are released to disseminate throughout the body. They can then attach to any "free" versions of the material that induced the B-cell response. This in itself can neutralize an infectious agent. For example, the antibodies coating a virus may physically block the virus' ability to coax a potential host cell from letting it in. Antibodies are also able to alert other arms of the immune system to assist, but those details will be revealed later in Chap. 13.

Following an encounter with material that their TCR and BCR receptors can recognize, some B and T cells will remain as long-lived memory cells. This allows your immune system to adapt to be prepared for what is now known to specifically exist in your environment and cause harm. This pool of memory cells is larger than the originally activated cells and can act more quickly and robustly. If the same material invades a second time, it is often disabled before it can cause noticeable symptoms. Creating memory cells is the purpose of vaccinations, which include inactivated or fragments of infectious agents. If an active version is encountered later, it can be promptly suppressed.

Unfortunately, for a large number of people, the immune system sometimes does attack self, resulting in an autoimmune condition. Almost 80 different types of autoimmunity have been described, including type I diabetes, multiple sclerosis, rheumatoid arthritis, and systemic lupus erythematosus (SLE). In most cases, autoimmune diseases are determined by the presence of antibodies to self (termed autoantibodies) and respond to treatments that mute the immune response. Rheumatoid factor (RF) was the first autoantibody identified in patients with rheumatoid arthritis. Intriguingly, RF is an antibody that binds to other antibodies and is found to be elevated in other autoimmune conditions too, such as SLE. In rheumatoid arthritis, autoantibodies aggregate in the fluid between the joints and lead to chronic inflammation, which leads to the damage of bone and cartilage. Autoantibodies to proteins specific to insulin-producing beta cells in the pancreas, which we learned about in the last chapter, are found in people with type I diabetes. Cytotoxic T cells are engaged to target the beta cells for apoptosis, leading to a life-long need for insulin injections. Treatments to suppress the immune system in autoimmune diseases obviously make people highly susceptible to infectious diseases. Thus, finding ways to only target the rogue components is highly desirable. Approaches to suppress only the subset of immune cells that improperly target self are looking promising; for example, silencing T cells that attack the myelin found on neurons (Chap. 7) in patients with multiple sclerosis is in clinical trials.

The ability of our immune system to detect non-self is a huge barrier to transplanting tissues or organs. The need to "match" the organ donor and the recipient is attributable to the MHCI receptors on cells (the exception is our enucleated RBCs, which do not express them). One person can have up to six different versions of genes encoding this receptor. The different versions of the receptors vary in what they can display to the immune system. If a patient receives an organ from someone with different MHCI receptors, it can display a fragment of a resident protein that will look foreign to the recipient's immune system and can culminate in organ rejection. Long wait lists for organ donations reflect not only the limited number of organs available but also the restriction of finding this genetic match. One possibility to meet this need is to use animals such as pigs as the source of organs. This should seem counterintuitive – certainly, pig hearts are seen as foreign by our immune system – the procedure is called *xeno*transplantation after all. However, the idea of "humanizing" animals has been in the works for a while, with numerous mouse models containing some human genes or cells well established as a means to study human diseases. Typically, humanized animals also have their immune system compromised so as to not reject the introduced human bits. Pig organs are much closer in size to humans than mouse organs, and with the recent advances in gene editing to humanize pigs, a handful of groundbreaking transplants were performed in 2021. There remains a lot to determine about the long-term success of this approach to trick the immune system and redefine what is "self."

11

Oocytes

From an evolutionary point of view, the query of *which came first, the chicken or the egg?* is straightforward to answer. Organisms have been reproducing using eggs for hundreds of millions of years before chickens arrived on the scene. Additionally, sexual reproduction (the key function of eggs) was established before any multicellular organisms existed. The immature form of an egg cell is called an oocyte. Unlike other cells, mature oocytes of animals are large enough to be seen by the unaided eye. Human oocytes measure about 100 microns in diameter, about the width of copy paper. Eggs that are laid, like chicken eggs, are especially large because packaged within the shell is not only the oocyte but also nutrient-rich yolk needed to support the growth of the embryo outside the mother's body or to support the diet of anyone who eats such eggs.

As cells used for sexual reproduction, oocytes need to deliver only about half of the genetic information of the parent to the offspring (it is a smidge more than half, as explained later in this chapter). This way, following fertilization, the total amount of deoxyribonucleic acid (DNA) remains the same from generation to generation. Dividing up DNA in this way occurs through a process called *meiosis*, with visibly distinct steps that many of us have been asked to memorize since primary school. Long before DNA was known to be the source of genetic information, images captured using microscopy revealed that chromosomes can align and separate with the help of cytoskeletal structures that form spindle poles and fibers (Fig. 11.1) as a cell divides into two. These structures are formed by microtubules, and not surprisingly, given that something inside the cell is moving in a purposeful manner, motor proteins are involved. We previously saw that microtubules can serve as a highway for

L. Saucedo, *Getting to Know Your Cells*, https://doi.org/10.1007/978-3-031-30146-9_11

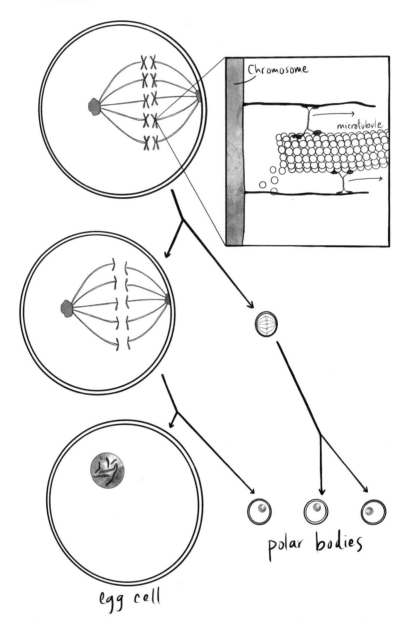

Fig. 11.1 *Meiosis*. The formation of an oocyte requires two rounds of division, which will result in the egg cell containing just one copy of each initial chromosome pair. These cell divisions are also asymmetric, which allows the oocyte to receive most of the intercellular contents. The separation of chromosomes during this process is accomplished by motor proteins pulling on chromosomes while moving along shrinking microtubules toward the cell periphery

neurotransmitter delivery (Chap. 7). In that scenario, they are stabilized by other molecules. But like microfilaments, microtubules can grow and shrink, and that ability is integral to allowing them to pull and push on chromosomes. However, they do not hold onto chromosomes directly. That is the job of motor proteins, which travel along the microtubules as their tails make attachments to regions on chromosomes, called *centromeres*. When chromosomes are being pulled, the microtubule is shrinking while the motor protein is moving away from the shrinking end (Fig. 11.1). It might help to imagine an action film where a bridge is crumbling apart at one end (the microtubule) while an actor races toward the other (the motor protein) carrying something important (the chromosome) with them. How DNA both aligns and separates in meiosis differs dramatically from *mitosis*, which leads to the duplication of virtually[1] genetically identical cells for organismal growth and maintenance. Importantly, it is not a matter of simply halving the DNA but rather carefully orchestrating which half of the DNA remains in the oocyte.

Like the vast majority of mammals, humans are *diploid*. This term reveals that we have two copies of each of our 23 chromosomes, for a total of 46. When oocytes halve their DNA, they become *haploid* cells, carrying just one copy of each chromosome. The two copies of each chromosome are considered *homologous*, containing instructions for the same genes, although frequently different versions or *alleles* of those genes. For example, chromosome 7 in humans contains the gene that encodes the protein cystic fibrosis transmembrane conductance regulator (CFTR), which transports an ion called chloride (Cl^-) across the plasma membrane. The function of this protein is especially important in the epithelial cells of the lungs. We know this because some people have versions of CFTR that malfunction, and this leads to cystic fibrosis; the buildup of chloride in cells disrupts the ability of airways to produce secretions that help clear away bacteria. All versions of CFTR are located in the same position on chromosome 7, and you inherited two: one from each biological parent. As long as one of your parents transmitted a functional version to you, you will not have cystic fibrosis.

Hopefully, you can appreciate why it is important that when you receive half of your DNA from a parent, each chromosome is included. For example, half of a parent's DNA could mean two copies of one chromosome (say chromosome 1) and no copy of another (say chromosome 7). In the vast majority of cases, a human embryo having fewer or more than two copies of a

---

[1] *Since DNA replication is not error-proof, some changes in DNA sequence will occur as described in Chap. 3.

chromosome (termed aneuploidy) is usually lethal (chromosome 21, and the X and Y chromosomes are exceptions). This is because the "dose" of many genes matters. The capacity to live with some variations of X and Y chromosomes is already in place – with XY and XX as the typical configurations. Chromosome 21 has the fewest number of genes, which may explain why having three copies is viable, although it does result in Down's syndrome. Interestingly, screening for aneuploidy as part of in vitro fertilization (IVF) has revealed that aneuploid cells that might arise by mitosis during embryogenesis (which takes place after meiosis and fertilization) can be eliminated by apoptosis, allowing embryos to develop into healthy babies.

While the process of meiosis carefully halves the DNA content of the resulting oocyte, it divides up the other materials of the cell asymmetrically. Meiosis involves one round of DNA replication, followed by two cell divisions, leading to four cells. But only one of these cells will become a mature oocyte and is distinguished by being dramatically larger than the other three. This is because the oocyte received much more than half of the *cytoplasm* and the components within during both divisions (Fig. 11.1). This is possible because chromosomes align toward the periphery of the cell rather than the center prior to cell division. The process that guides the pinching of the plasma membrane into two cells during division coordinates with where chromosomes align (and is an additional means to ensure each cell gets half the DNA). The three cells that do not develop into an oocyte are called *polar bodies*. At a minimum, polar bodies serve as a repository of "excess DNA" not meant for the next generation and are actively destroyed in mammals. But in some animals, the fusion of a polar body with an oocyte is used for *parthenogenesis,* in which no sperm is involved in fertilization and the offspring has just one parent. Receiving more than half of the cytoplasm during oogenesis means the oocyte will contain surplus molecules and organelles. This is important because during fertilization, the sperm will contribute its half of DNA but very little cytoplasm. For example, during fertilization, the sperm contributes 1000-fold fewer mitochondria, and there are even mechanisms in place to degrade paternal mitochondria (as we will see in Chap. 12) such that it is very rare to find any intact after the first few cell divisions following the formation of the zygote. Mitochondria actually carry a very small amount of DNA (just 0.0005% of the total DNA). Thus, the oocyte actually contributes slightly more than half of the DNA to the next generation. While small in amount, mitochondria DNA encodes machinery that is essential in converting energy during aerobic respiration (Chap. 5). Diseases resulting from mutations in mitochondrial genes are inherited from the mother and can be severe.

Importantly, even though a zygote receives half of its DNA from each parent, the ability of a gene to be expressed can differ based on which parent it was inherited from. There are approximately 150 genes in humans that demonstrate parentally restricted expression, and genes with this relatively unusual additional layer of management are referred to as being *imprinted*. One such imprinted gene is called insulin-like growth factor 2 (IGF2). Upon fertilization, the gene product of IGF2 is expressed from the paternal chromosome but not the maternal chromosome. Importantly, this difference is not due to any variance in the DNA sequence of IGF2 but rather a result of the chemical modifications of DNA and DNA-associated proteins, which alter how accessible a gene is to be read by RNA polymerase (Chap. 3). Chemical modification of IGF2 in the oocyte results in the maternal copy of IGF2 being "silenced" during embryogenesis. Intriguingly, IGF2 promotes the growth of the fetus, and so it has been postulated that the silencing of IGF2 on the maternal chromosome helps restrict the size of the offspring, which needs to pass through the tight space of the birth canal, thus improving the chances of the mother surviving childbirth. In contrast, the unhindered expression of IGF2 from the paternal chromosome should promote offspring size and thus boost its chances of survival. Studies using genetically modified mice support this model; offspring without are paternal copy of IGF2 are 40% smaller. The importance of imprinted genes became evident when the technique of cloning animals became prevalent. Cloning utilizes DNA from just one parent and from cells that do not undergo meiosis. Many of the growth and health abnormalities seen in cloned animals appear to be due to bypassing the process of imprinting genes, which normally occurs during the maturation of oocytes (and sperm).

Some human disorders have been tied to imprinted genes. Interestingly, the deletion of a small region of chromosome 15 leads to different outcomes depending on the parent it is inherited from. If the deletion is on the paternal chromosome, Prader-Willi syndrome occurs, but if the deletion is located on the maternal chromosome, then Angelman syndrome arises. While the syndromes have some overlapping features, they are easily clinically distinguishable. The explanation for why the outcome differs based on what parent the deletion is inherited from reflects that some of the genes located in that region of the chromosome are only expressed by the paternal chromosome (lost in Prader-Willi), and other genes at that location are only expressed by the maternal chromosome (lost in Angelman syndrome).

It is important to recognize the impact of genetic changes or errors in the oocyte (or sperm). Unlike alterations in the other cell types discussed in this book, they will be propagated via mitosis to every cell of the organism that develops and, also, to the next generation.

12

Sperm

Possibly the most recognizable of all cell types, sperms are unique among mammalian cells for their long "tail" or *flagellum*, which allows them to swim (Fig. 12.1). The "head" or cell body of sperm is quite small; at about 3 micrometers, its width is less than half that of a red blood cell. And yet as one of the first cells to be observed using microscopy during the 1600s, scientists proposed that a miniaturized human was located within this space. This concept was called "preformation," and following fertilization, it was proposed that the human simply grew larger. Instead, it is now understood that the cell body of sperm contains half the nuclear deoxyribonucleic acid (DNA) for the next generation and that the rest of sperm structures exist to empower sperm to deliver this DNA.

Similar to oocytes, sperms undergo meiosis as they develop. Prior to differentiation, early *spermatids* are round cells but then undergo dramatic remodeling, which includes eliminating most of the cytoplasm and organelles other than the nucleus while also forming the trademark tail. The extrusion of cytoplasm and organelles into what is termed a cytoplasmic droplet from developing sperm is important as too much "residual cytoplasm" is associated with infertility. During remodeling, an organelle unique to sperm, called the *acrosome*, is assembled. It appears to arise from vesicles released from the Golgi apparatus (prior to its expulsion), which fuse together into a larger structure situated above the nucleus at the front tip of sperm. The interior of the acrosome is acidic and contains enzymes that are key to breaking through the protective layer surrounding the oocyte during fertilization. The tail of sperm

L. Saucedo, *Getting to Know Your Cells*, https://doi.org/10.1007/978-3-031-30146-9_12

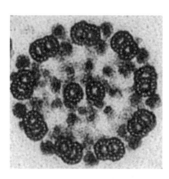

Fig. 12.1 *Sperm and flagellum.* Electron microscopy images of intact sperm (top, adapted from https://onlinelibrary.wiley.com/doi/abs/10.1111/and.13043) and a cross section of the flagellum revealing doublets of microtubules contributing to the axoneme structure (bottom, adapted from https://link.springer.com/article/10.1007/s10815-016-0652-1)

will form opposite of the location of acrosome, starting as a small protrusion that elongates into a flagellum, which is typically ten times the length of the cell body (Fig. 12.1).

While cell crawling predominantly uses microfilaments to push a membrane forward (Chap. 6), cell swimming via the flagellum utilizes microtubules organized into a complex structure, called the *axoneme*, to propel the cell (Fig. 12.1). In fluid, sperms travel about 5 millimeters/min, which is about 5000x faster than a crawling fibroblast. We previously saw that a microtubule can serve as a highway for motor proteins carrying cargo (Chap. 7) and as a means to pull chromosomes (Chap. 11). But in the context of the axoneme, microtubules are instead paired up to form doublets. These doublets are arranged with nine evenly spaced near the plasma membrane, and two at the center of the flagellum. This organization is visible when viewing a cross-section of a flagellum (Fig. 12.1). Several other structures exist to connect and stabilize the microtubule organization in the axoneme, but similar to what we saw for muscle contraction (Chap. 8), an immobilized motor protein plays a key role in moving the cell. In this case, the motor protein is called *dynein*. In contexts other than the axoneme, dyneins can carry cargo while using their head domains to walk along microtubules. But in the axoneme, the part of

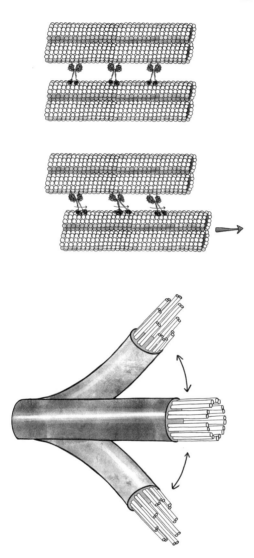

Fig. 12.2 *Flagellar motion.* Dynein motor proteins make connections between the neighboring doublets of microtubules (top). Using energy released from ATP, the immobilized dyneins slide the doublets relative to one another. This happens in a coordinated manner between all nine outer doublets of the axoneme, causing the flagella to move and enabling sperm to swim (bottom)

dynein that could bind to cargo is instead securely attached to a microtubule, forming one of the outer doublets. Meanwhile, the head domain of the dynein is positioned to interact with a neighboring doublet (Fig. 12.2). When the energy from adenosine triphosphate (ATP) is converted to cause movement of the head domains, they are able to bind to and slide the neighboring doublet, leading the flagellum to bend (Fig. 12.2). Coordination of the sliding movement among all nine doublets leads the tail to exhibit a corkscrew motion. In humans, sperms need to travel 15 or more centimeters. This requires a tremendous amount of ATP, and so dozens of mitochondria are packed into the flagellum, in a region called the midpiece, positioned below the sperm head like a turtleneck.

Sperms need to conserve their energy until fertilization is a possibility, and so it makes sense that tail movement is low until an activation signal is received. This activation begins when sperms encounter a higher concentration of Ca^{2+} in the female reproductive tract, though the mechanism between Ca^{2+} levels and dynein activity is not clear. Importantly, swimming faster is pointless unless heading in the right direction, and so sperms likely have the ability to sense the location of the oocyte. Some data indicate that the hormone progesterone, which is secreted by cells that surround an ovulated oocyte, may help orient sperm. Sperms have progesterone receptors to sense its presence; however, progesterone affects sperm in multiple ways that make it difficult to sort out each contribution to sperm success. For example, progesterone causes intracellular Ca^{2+} to rise even more in sperm, leading to the hyperactivation of tail movement. Progesterone also helps sperms to prepare their acrosome to be competent to fuse with the sperm membrane so as to release its contents via exocytosis upon reaching the oocyte.

Of the approximately 100,000,000+ sperm released during human ejaculation, only a couple hundred are thought to reach the oocyte. It is important that not more than one sperm fertilizes the egg due to the lethality of aneuploidy, as we learned about in the last chapter, and so following the penetration of the oocyte's plasma membrane, the egg secretes molecules that are resistant to the degradative enzymes within the acrosome of subsequent sperm (while not of a mammal, a video using sea urchin eggs and sperm can be found at https://www.youtube.com/watch?v=fO4UWj01Gx8 and captures the formation of a fertilization envelope to prohibit a second sperm from penetrating an already fertilized egg).

While it is only the nucleus that needs to be deposited within the oocyte, the midpiece of the sperm and all its mitochondria also enter. Mitochondria contain just a few dozen genes, but mechanisms for offspring to inherit mitochondria from only one parent have been strongly selected during evolution. In mammals, it is the mitochondria from the egg that pass on to the next generation. Why this is the case is unclear, but mice that were engineered to contain two sets of mitochondria had impaired energetics, behaviors, and cognition. Multiple mechanisms have been suggested to ensure that sperm mitochondria are incapacitated, including a process called *autophagy*. Autophagy is a means by which cells break down their own organelles, typically when they are damaged or when a cell is so depleted of energy resources that it uses this process as a way to free up the building blocks needed to survive. During autophagy, the organelle to be degraded is surrounded by a membrane, which then fuses with another organelle called the *lysosome*. Lysosomes are acidic and contain enzymes to digest complex molecules into their basic units, which are then recycled. Mitochondria from sperm appear to be marked for autophagy prior to fertilization but do not encounter the machinery to act on it until it is inside the organelle-rich oocyte. Unlike genes encoded by nuclear DNA that we receive from both parents inheriting just one copy of each gene that is encoded in mitochondrial DNA means there is no "back up version" in case of mutation. Because inherited mutations in mitochondria can disrupt the main means of producing ATP, the outcome is usually severe or lethal. To circumvent this outcome in families with known risk, mitochondria from a donor are used to replace the mitochondria of an oocyte before utilizing in vitro fertilization (IVF). This is how "three-parent" children have been born.

Sperms are the one cell type with a purpose that resides outside of the organism that produces them. Maybe this distinction partially accounts for why they are perhaps the only cell type to have been depicted by actors and the inspiration for the lyrics of at least a dozen songs, often about the "race" to reach an oocyte. But it also means that a cellular objective can be prevented by using physical barriers that impede sperm from reaching and fertilizing eggs. Inhibiting sperm function via medicinal approaches is also a possibility. Because sperms are unique among human cells in both having a flagellum and an acrosome, these structures are the obvious targets to impair. However, one limitation is that the cellular projections called cilia, which we encountered in Chap. 4, share mechanics with flagella and are found on numerous cell types. Cilia are shorter than flagella, and rather than move the cell, they move material across the cell surface; for example, respiratory epithelial cells move mucus from our lungs up into our throat (which then we either swallow or cough out). The same axoneme structure is required for this function, so it would be dangerous to develop drugs that indiscriminately prevent its motion. Therefore, finding pharmaceuticals that impede the activation step of sperm mobility or the formation of the acrosome would be more promising as birth control with minimal side effects.

13

The Immune System

We got an introduction to a subset of white blood cells that are able to use cellular interactions to develop the ability to tell self from non-self in Chap. 10. In this chapter, we will take a broader perspective to see how several more cell types communicate and cooperate to establish a complex and robust system to protect us from unwelcomed elements from our environment and perhaps also expand the demarcation of what we consider to be self.

As a reminder, our epithelial cells are our first-line defense between us and our surroundings. As we saw in Chap. 6, disruption of the dense packing of epithelial cells triggers a wound response in the underlying tissue. Part of this response includes recruiting immune cells to the area to search for any possible invaders while the tissue repair proceeds. Damaged epithelial cells release molecules from their cytoplasm, collectively called "danger-associated molecular patterns" (DAMPs), which can attract our most abundant type of white blood cell called a *neutrophil*. If the wound damaged a blood vessel, neutrophils are able to exit circulation with ease and remain in the region, thanks to the binding of plasma membrane receptors to the DAMPs (Fig. 13.1). But even if blood vessels are not directly damaged, a nearby wound can trigger them to leak. Blood vessels are lined by a specialized type of simple, squamous epithelial cell called an *endothelial cell*. Normally, the tight junctions between endothelial cells prevent fluid and cells within blood vessels from exiting into nearby tissue. But endothelial cells are also able to bind to DAMPs, and in turn, this leads them to augment the immune response. They release additional molecules called *cytokines*, which continue to attract immune cells and assist them in their various functions. These cytokines also help dismantle the tight junctions between endothelial cells, allowing both immune cells and

L. Saucedo, *Getting to Know Your Cells*, https://doi.org/10.1007/978-3-031-30146-9_13

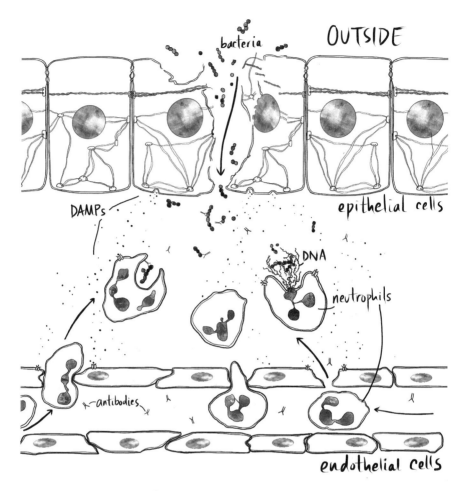

Fig. 13.1 *Cells working together to mount an immune response.* Wounding of the epithelial cell barrier sets into motion numerous reactions to protect against the infiltration of potential pathogens, like bacteria. Epithelial cells release DAMPs that signal to endothelial cells lining blood vessels and neutrophils. This causes the vessels to become leaky and attracts neutrophils to exit vessels at the wound site. Neutrophils can help trap pathogens by expelling their DNA and can also digest them via phagocytosis. If the pathogen has been previously encountered, B cells (not shown) can enhance the efficiency of neutrophil function by tagging the pathogens with the antibodies they secrete

fluid into the nearby tissue. We experience this as localized swelling. It may seem counterintuitive to have compromised blood vessels near a wound since that might enable external material to gain access to our entire body. But the hydrostatic pressure created by the leakage can physically push material away from the vessel and trap it near the wound boundary. Altogether, this response to a wound is what we call inflammation and happens indiscriminately regardless of what may have breached our defenses. It is why we might feel ill following vaccination, even when it contains no intact pathogens; many of the symptoms that we associate with being sick are from an effective immune reaction and not what triggered it. This type of immune response is referred to as "innate" because all the participating mechanisms are encoded in the genes that we have inherited and are not directed toward a specific target.

Neutrophils that arrive on the scene are able to engulf and digest any bacteria or viruses through a process called *phagocytosis*. Here, microfilaments push the plasma membrane of the neutrophils toward the target and, when close, extend thin projections from the membrane to surround it. Whatever is captured is now enclosed within a compartment called a phagosome. It will next fuse with a lysosome, leading to the degradation of its contents (a remarkable video of a neutrophil chasing bacteria among red blood cells can be seen here: https://www.youtube.com/watch?v=I_xh-bkiv_c). Neutrophils can also secrete small antimicrobial molecules that can bind and disrupt the plasma membrane or cell wall of bacteria. Our hepatocytes are continuously producing and secreting these antimicrobials into our bloodstream too (as well as the clotting factors that help stop blood loss from badly damaged vessels). And finally, neutrophils can release fibrous traps, which include their own DNA, and entangle nearby microbes (Fig. 13.1).

Of course, many pathogenic bacteria and viruses have evolved to counter the attacks of neutrophils, such as changing their surface molecules to evade phagocytosis and antimicrobial molecules as well as secreting enzymes to break down the extracellular DNA component of traps. And so additional rounds of a more targeted defense involving the B and T cells we encountered in Chap. 10 will be crucial. Importantly, the inflammation needs to subside once the wound is closed because our own cells can also be damaged by the response, and the vasculature needs to reseal. Chronic inflammation is associated with many diseases, including cardiovascular and autoimmune diseases. Though underlying reasons vary, a cycle of damaged cells releasing DAMPs leads to neutrophils damaging even more cells. Avoiding this sequence of events is likely why the apoptotic version of cellular death that we encountered in Chap. 10 avoids allowing cytoplasmic contents (including DAMPs) from spilling out into the environment. The carefully orchestrated steps of apoptosis ensure that cellular debris remains in membrane-bound vesicles, which are cleared up via phagocytosis without eliciting an inflammatory response.

Importantly, our immune system is not dependent on a wound to prompt a defensive call to action. Some immune cells are instead taking offensive measures. *Dendritic cells* are a type of white blood cell that is especially skilled at patrolling for possible stealth intruders. These cells constantly travel to the epithelium of the skin, lungs, and guts to survey the surrounding environment. If they encounter something with a signature that they can recognize as potentially harmful, they will capture it by phagocytosis. However, instead of fully destroying it, they will use the secretory pathway (Chap. 6) to display small components of the encapsulated material on plasma membrane receptors. The dendritic cell can then travel to a lymph node to present what it found to T cells in a less rushed setting than the circulatory system. This will then lead to the activation of T cells and B cells, as described in Chap. 10. In this way, dendritic cells fulfill their job as "antigen-presenting cells" (APCs). An *antigen* is anything that can lead to *anti*body *gen*eration by B cells. However, one difference is that unlike when a T cell encounters an infected host cell, it need not kill the dendritic cell. This is because the dendritic cell actively captures the suspect material and keeps it enclosed in a membrane during phagocytosis and secretion to send it back to the plasma membrane. This means that ensnared bacteria or viruses never gain access to the cytoplasm or nucleus and so cannot use the cellular machinery needed to replicate themselves. Like neutrophils, dendritic cells are part of our innate immune system – they only recognize generic features of some pathogens that remain static over time. But following phagocytosis, they are able to display any random part of the digested pathogen, including bits that innate cells cannot recognize. This creates the opportunity to summon the "adaptive" part of our immune system, which we saw in Chap. 10: the T and B cells. They are termed adaptive because the random rearrangement of their DNA enables them to target new variations or completely novel pathogens. The adaptive and innate "arms" of our immune system work together in other ways. For example, antibodies produced by B cells (Chap. 10) bind not only to their specific target but also to neutrophils (Fig. 13.1). This stimulates their phagocytosis and that of the pathogen they are attached to. In this way, the adaptive immune system helps mark trespassers that the innate immune system may not be able to detect, and it is especially important for the immune system to stay on top of pathogens, which can evolve quickly.

Although pathogenic bacteria get the most press, most bacteria likely do us no harm, and many assist us. These types of bacteria are called "commensal," and they are highly abundant on our skin and in our gut, with a total number about equivalent to the number of cells we would call our own. But since they are generally much smaller in size, their mass only equates to less than a half-pound of an adult's weight. Coexisting with these microbes is important to avoid constant inflammation and to derive benefits from them. This is especially true in the gut, where bacteria help us digest food, supply us with vitamins, and secrete inhibitory molecules that keep their harmful relatives at bay. Research shows that T cells that can recognize beneficial bacteria are in circulation, but importantly, this includes a class of cells called regulatory T cells or Tregs. Tregs suppress inappropriate actions by the immune response, such as autoimmunity and allergic responses, and also keep appropriate reactions from becoming so extreme that they cause more harm than good. Dendritic cells that reside in the gut are able to coax immature T cells to differentiate into Tregs. Tregs then secrete compounds that stifle innate immune cells from causing inflammation and keep cytotoxic T cells and B cells from mounting a response. While not formally "us," our resident microbes contribute to our best selves, which is easy to appreciate after a round of antibiotics wipes them out and wreaks havoc on the health of our digestive system. Attention to hosting bacteria that are optimal in our gut has grown tremendously in the past 20 years, including an emphasis on probiotics in food and supplements to the much more dramatic use of fecal transfer for more severe intestinal problems.

While many microorganisms can be neutral or beneficial, enough of them are a threat such that living with a compromised immune system is extremely challenging. Some humans have been born with a genetic condition that profoundly compromises the adaptive arm of the immune response, called SCID (severe combined immunodeficiency). Without functional B cells and T cells, these patients are unlikely to see their first birthday unless kept in a sterile environment until successfully treated with a transplant from a healthy, matching donor to repopulate the stem cells of their blood. Numerous people who need to actively suppress their immune systems (organ recipients, those with human immunodeficiency virus (HIV) or debilitating autoimmune conditions) are much more susceptible to infections and are less able to notice they are infected since so many of the symptoms we experience are due to the inflammatory response.

It is crucial for our immune system to not attack self and our welcomed inhabitants. However, there is a strong incentive to carefully train it to detect a subset of our own cells: those that have become cancerous. We will see in the next chapter how this can work.

14

Cancer

An individual cell that is not able to perform its function is usually not a call for concern as there are millions of other healthy cells that can easily compensate. It is more alarming when a cell abuses its function, as we have seen with autoimmunity. However, when malfunctions empower a cell to disregard the rules in place to safeguard the success of a multicellular organism, the situation becomes especially dangerous. Cancer is the term to capture a large set of diseases characterized by inappropriate and unrestricted cellular behaviors. It requires several genetic mutations to occur in at least one cell, called a "founder." While one can inherit genetic predispositions to cancers, most genetic mutations are acquired during one's lifetime. In short, these mutations help counteract numerous protective mechanisms, such as preventing apoptosis and activating telomerase, which enable cells to multiply beyond their normal limits at the expense of the patient.

While cancers are usually named according to the tissue they arise from, the specific cell type that is the founder of the vast majority of cancers is the epithelial cell. One reason for this is that epithelial cells experience the most direct environmental insults, including damage to deoxyribonucleic acid (DNA). Another reason is that epithelial cells have a relatively high rate of division, and every time a cell divides, enzymes that create new DNA introduce errors. Sequencing the DNA of over a dozen different cancer types has indicated that about 2/3rd of genetic alterations are due to DNA replication errors. This is likely why age is the biggest risk factor for cancer; the cells of older people have simply divided more times and thus accumulated and perpetuated more genetic errors. For a long time, cancer research focused only on the cancer cell, a rogue that became self-sufficient by making its own growth

L. Saucedo, *Getting to Know Your Cells*, https://doi.org/10.1007/978-3-031-30146-9_14

factors and ignoring inhibitory signals from neighboring cells that are getting crowded out. But cancer cells often benefit by manipulating the skills of other cells. While epithelial cells are the cells that are most likely to become cancerous, they get significant assistance from several other cell types, including fibroblasts and immune cells.

As you recall, in serving their protective function, epithelial cells need to be packed together tightly and even use proteins in their membranes to hold onto one another. In order for a cell to become cancerous, these cell-to-cell interactions need to be severed. A key protein that is disrupted in cancer cells is called *E-cadherin* (E for "epithelial"). This protein spans the plasma membrane, and its extracellular portion binds to another E-cadherin on neighboring cells, forming an *adherens junction* (Fig. 14.1), which is located below the tight junctions that we saw in Chap. 4. Additionally, the intracellular side of E-cadherin interacts indirectly with microfilaments. The loss of E-cadherin ends up promoting cancer in multiple ways. First, it allows neighboring cells to break one of the connections they use to hold on to one another. Then by freeing up the microfilaments that were secured to the adherens junction, these microfilaments are now able to instead engage in cell crawling and contribute to the final step of cell division when the cell is cleaved in half, called *cytokinesis* (Fig. 14.1). Staining for E-cadherin in cancer biopsies is used to

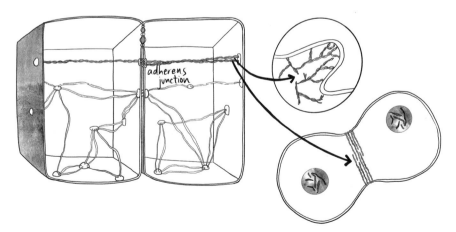

Fig. 14.1 *Epithelial cells with adherens junctions, intact or not.* Adherens junctions assist in holding neighboring epithelial cells together while simultaneously interacting with and organizing microfilaments inside cells (left). In cancer, the loss of adherens junctions occurs if a protein called E- cadherin is not produced. This allows epithelial cells to separate from one another and release microfilaments, which are now available to help cells crawl and divide (right)

predict prognosis in multiple types of cancer, such as breast, bladder, and prostate. In most cases, less E-cadherin is indicative of more invasive cancer.

In a sense, as epithelial cells convert into *carcinomas*, they begin to act more like unconstrained fibroblasts. This process is referred to as *epithelial-to-mesenchymal transition (EMT)*. They even start to produce proteins that normally serve as markers for fibroblasts, such as a cytoskeletal protein called vimentin. Normally packed together tightly and often partitioned from underlying tissues, rogue epithelial cells on the move also need to create a path for themselves. This can be accomplished by secreting molecules that break down existing ECM. Intriguingly, rather than secrete those degradative molecules themselves, cancerous epithelial cells can entice noncancerous fibroblasts to take on that task! These assistants are called *cancer-associated fibroblasts* (CAFs).

As cancer proceeds, the cells invade neighboring tissues and can even enter the bloodstream to travel to and colonize a secondary site. This process is called *metastasis*, and once cancer has metastasized, a patient's prognosis significantly worsens. The metastasis of cancer cells is greatly aided by both CAFs and immune cells (Fig. 14.2). Not only do CAFs help create physical pathways for cancerous epithelial cells to leave their home tissue, but they can also help prepare a welcoming environment in a new tissue for cancer to settle. Normally, cells from one tissue are not able to reside in another due to a lack of appropriate ECM and growth factors. CAFs are able to send signals through the bloodstream to the secondary site to establish a supportive microenvironment before the cancer cells arrive. Sending signals through the bloodstream to a different tissue is a much easier feat than sending actual cells, which are much larger than signaling proteins or vesicles.

Cancerous epithelial cells must both enter and exit the bloodstream to travel to a second site. As we learned in the preceding chapter, blood vessels are lined with endothelial cells bound together by tight junctions to prevent unregulated passage between them. But we have also learned that some immune cells are granted access across endothelial cells so they can exit and reenter the bloodstream as they search for intruders. Cancerous epithelial cells exploit this capability of immune cells. In some ways, a developing cancer is similar to a wound because the normal architecture of the tissue becomes distorted. This triggers multiple aspects of the immune response, including inflammation and the recruitment of a category of white blood cells, called macrophages, to the region. Inflammation loosens the tight junctions between endothelial cells, allowing macrophages to move across vessels at the site of cancer. Similar to the process of fibroblasts being co-opted by cancer cells, the macrophages taking part in this process are given a qualifying label: *tumor-associated macrophages (TAMs)*. In most cases, high numbers of TAMs

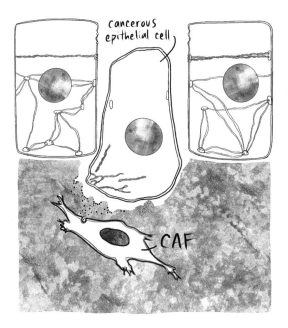

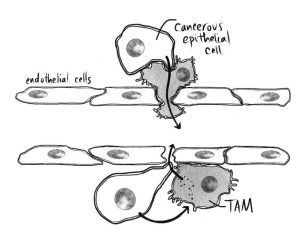

Fig. 14.2 *Role of fibroblasts and macrophages in supporting cancer.* Cancerous epithelial cells signal to nearby fibroblasts (cancer-associated fibroblasts (CAFs)) and entice them to assist cancer progression (top). Here, one CAF is secreting degrative enzymes to break down the extracellular matrix (ECM) so the cancer cell can crawl into underlying tissue. Tumor-associated macrophages (TAMs) can also promote cancer progression by facilitating the ability of cancer cells to enter (and exit) the bloodstream (bottom). Either cancer cells can piggyback onto TAMs as they cross the endothelial cell barrier, or cancer cells can elicit signals from TAMs to allow them to cross directly

correlate with a poorer prognosis. While TAMs have been implicated in multiple aspects of cancer, during metastasis, they appear to be rate limiting for cancer cell entry (intravasation) and exit (extravasation) from the bloodstream. The removal of a subset of macrophages in genetically engineered mice strongly suppressed the ability of breast cancer cells to form metastases in the lung. While it appears that cancer cells can simply follow the route of a macrophage, many studies show direct signaling between the two cell types, which empowers the cancer cells to cross the endothelial barrier (Fig. 14.2).

The idea that the immune system should fight cancer (rather than aid it) has garnered a lot of interest. But that means the immune system needs to "see" cancer cells as foreign even though they are our own cells. As discussed in Chap. 10, the immune system has safeguards in place to prevent attacks of self, and when it fails, numerous autoimmune disorders can arise. Much evidence points to the immune system not being very effective at combating cancers unless they are among the handful that are caused by viruses (for example, the causative role of human papillomavirus (HPV) in cervical cancer). In those cases, it is the viral proteins being produced within your cells that summon the immune response. However, cancer cells that are not infected with viruses can also put immune cells on alert. This is because cancer cells contain genetic mutations, which can give rise to proteins that do not otherwise exist in a patient and therefore can serve as a beacon to identify a cancer cell as unusual. There is evidence that this happens; a process called *immunoediting* has been observed, in which cancer adjusts its characteristics following recognition by immune cells. But it appears that most cancers successfully escape this endeavor by the immune system to contain them. Nonetheless, *immunotherapy* is an approach that is under very active investigation in hopes to improve the ability of the immune system to eliminate cancerous cells while ignoring healthy ones. One approach is to remove and genetically modify a patient's T cells to recognize a specific protein altered in their cancer, then reintroduce it. At the forefront of this field is the treatment of melanoma, with significant success in prolonging the lives of patients.

One issue that will continue to preclude curing cancers is that cancers are usually not detected until they have produced hundreds of thousands to millions of cells. These cells are also typically genetically unstable, so they continue to evolve quickly. Effective treatment can eradicate 99% of cancer, but the cells left behind generally are those that are resistant to the treatment and continue to grow until they are detected months or years later. Because of this, much research focuses not on the cancer cells themselves but instead on the healthy cells, such as the CAFs and TAMs discussed here. Because they are genetically stable, targeting them so that they can no longer assist cancer cells may prove valuable.

It may feel frustrating that cancer remains such a deadly disease over 50 years after President Nixon declared "war" against it. At the time, previous successes with infectious diseases inspired confidence. However, unlike infectious disease, each person's cancer is uniquely their own, arising from their cells with a unique genetic signature. Many modern treatments have therefore moved toward a personalized medicine approach. Newer chemotherapies are far more directed, specifically inhibiting or eradicating cells with mutated proteins known to contribute to cancer, as we saw with tamoxifen and ER+ tumors in Chap. 9. This is in strong contrast to classical approaches that targeted all highly proliferating cells, leading to very severe side effects by destroying healthy cells, notably epithelial cells that line the digestive system and hair follicles, as well as white blood cells of the immune system.

One thing to keep in mind is that while cancer is so prevalent among people, at the cellular level, it is incredibly rare. Although almost one in every two people will develop cancer in their lifetime, that detectable cancer represents one founder cell out of the approximately 37 trillion cells in a person who was able to overcome all the barriers in place to prevent it. Improving upon that rate of prevention is a daunting but worthy goal.

15

Conclusion

The seventeenth-century poet John Donne's contention that "no man is an island" was inspiring in the context of underscoring the interconnectedness between people. But at the cellular level, it is my hope that you now appreciate how we each could indeed be viewed as an island, populated by dozens of cell types interconnected by shared DNA to perform their individual contributions to the monumental task of defining and supporting our existence. Some cell types are tight-knit, defining the barrier between us and our environment, while others just go with the flow, via our circulatory system. I hope I have also been able to convey just how dynamic most of these cells are, whether crawling within your dermis or having cargo motoring down a cytoskeletal highway within them.

Multicellularity has allowed cells to become experts in their fields – be it nerve transmission or hunting down trespassers. But that only works if they remain part of the island; none are self-sustaining to "sail off" on their own. It also means that each cell type is usually only reacting to what it needs to perform its role. In essence, by only expressing a subset of genes, each cell type is restricting its capacity to do it all. In contrast, unicellular organisms can seemingly exist independently and thus give the impression that they do do it all. But really, they do just what they must to reproduce, just like all cells. Along this line of thinking, in her book "Pilgrim at Tinker Creek," Annie Dillard laments that her brain edits what she is able to sense from the world around her and quotes Donald E. Carr, who said "… that only the simplest animals perceive the universe as it is." But this misses the point that the default state is to not perceive – that it takes a molecule (usually a protein encoded by a gene) to receive the information. The "simpler" a cell is, the less likely it is to be

L. Saucedo, *Getting to Know Your Cells*, https://doi.org/10.1007/978-3-031-30146-9_15

devoting resources to sensing inessential aspects of its environment. While it would be interesting to "force" a cell from a multicellular organism to simultaneously express its full arsenal of genes, it seems likely that it would result in cellular chaos rather than an omniscient cell.

As multicellular organisms, we get the benefit of starting each day at least partially anew. For us, the phrase "lose yourself" can be taken literally! Over 300 billion cells are eliminated and replaced each day in the human body (about 1% of our total cells). I, for one, am glad that those cells are mostly invisible to the eye and are dispersed throughout my body. It would be quite disturbing to experience even a very small chunk of tissue breaking off and regenerating on a daily basis. And our obliviousness to it certainly helps us maintain a static image of who we are. But it might be helpful to embrace the idea that at the cellular level, you are given new life every day: 300+ billion cells dedicated to refreshing the island of you.

Sources

Chapter 2

Ji P, Murata-Hori M, Lodish HF. Formation of mammalian erythrocytes: chromatin condensation and enucleation. *Trends Cell Biol.* 2011;21(7):409-415. doi:https://doi.org/10.1016/j.tcb.2011.04.003 *Describes process and purpose of enucleation of mammalian RBCs.*

Berg JM, Tymoczko JL, Stryer L. Biochemistry. 5th edition. New York: W H Freeman; 2002. *Describes binding site of CO2 and ability to facilitate release of O_2.*

Blumenthal I. Carbon Monoxide Poisoning. *Journal of the Royal Society of Medicine.* 2001;94(6):270-272. doi:https://doi.org/10.1177/014107680109400604

Hajdu, S. The Discovery of Blood Cells. *Ann Clin Lab Sci* Spring 2003; 3(2):237–238

André, Bruno, Virginie Raynal, Baya Chérif-Zahar, Jean-Pierre Cartron, Giorgio Matassi, and Anne-Marie Marini. "The Human Rhesus-associated RhAG Protein and a Kidney Homologue Promote Ammonium Transport in Yeast." *Nature Genetics* 26.3 (2000): 341-44. doi:https://doi.org/10.1038/81656

Gallagher, P. Red Cell Membrane Disorders. *Hematology Am Soc Hematol Educ Program* 2005; (1):13–18. https://doi.org/10.1182/asheducation-2005.1.13

Eva Bianconi, Allison Piovesan, Federica Facchin, Alina Beraudi, Raffaella Casadei, Flavia Frabetti, Lorenza Vitale, Maria Chiara Pelleri, Simone Tassani, Francesco Piva, Soledad Perez-Amodio, Pierluigi Strippoli & Silvia Canaider (2013) An estimation of the number of cells in the human body, *Annals of Human Biology*, 40:6, 463-471, DOI: https://doi.org/10.3109/03014460.2013.807878

Chapter 3

Yamanaka S, Li J, Kania G, et al. Pluripotency of embryonic stem cells. *Cell Tissue Res.* 2008;331(1):5-22. doi:https://doi.org/10.1007/s00441-007-0520-5

http://sitn.hms.harvard.edu/flash/2014/stem-cells-a-brief-history-and-outlook-2/

Appelbaum FR. Hematopoietic-cell transplantation at 50. *N Engl J Med.* 2007;357(15):1472–1475. doi:https://doi.org/10.1056/NEJMp078166

Atari M, Gil-Recio C, Fabregat M, et al. Dental pulp of the third molar: a new source of pluripotent-like stem cells. *J Cell Sci.* 2012;125(Pt 14):3343–3356. doi:https://doi.org/10.1242/jcs.096537

Baulies A, Angelis N, Li VSW. Hallmarks of intestinal stem cells. *Development.* 2020 Aug 3;147(15):dev182675. doi: https://doi.org/10.1242/dev.182675.

Bodnar AG, Ouellette M, Frolkis M, et al. Extension of life-span by introduction of telomerase into normal human cells. *Science.* 1998;279(5349):349–352. doi:https://doi.org/10.1126/science.279.5349.349

https://www.the-scientist.com/news-opinion/increasing-number-of-ips-cell-therapies-in-clinical-trials--65150

Boruczkowski D, Pujal JM, Zdolińska-Malinowska I. Autologous cord blood in children with cerebral palsy: a review. *Int J Mol Sci.* 2019;20(10):2433. Published 2019 May 16. doi:10.3390/ijms20102433

Cofano F, Boido M, Monticelli M, et al. Mesenchymal Stem Cells for Spinal Cord Injury: Current Options, Limitations, and Future of Cell Therapy. *Int J Mol Sci.* 2019;20(11):2698. Published 2019 May 31. doi:10.3390/ijms20112698

Bartolucci J, Verdugo FJ, González PL, et al. Safety and Efficacy of the Intravenous Infusion of Umbilical Cord Mesenchymal Stem Cells in Patients With Heart Failure: A Phase 1/2 Randomized Controlled Trial (RIMECARD Trial [Randomized Clinical Trial of Intravenous Infusion Umbilical Cord Mesenchymal Stem Cells on Cardiopathy]). *Circ Res.* 2017;121(10):1192–1204. doi:https://doi.org/10.1161/CIRCRESAHA.117.310712

Chapter 4

Peters B, Kirfel J, Büssow H, Vidal M, Magin TM. Complete cytolysis and neonatal lethality in keratin 5 knockout mice reveal its fundamental role in skin integrity and in epidermolysis bullosa simplex. *Mol Biol Cell.* 2001;12(6):1775–1789. doi:https://doi.org/10.1091/mbc.12.6.1775

Hou J. The kidney tight junction (Review). *Int J Mol Med.* 2014;34(6):1451–1457. doi:https://doi.org/10.3892/ijmm.2014.1955

Rescigno M, Urbano M, Valzasina B, et al. Dendritic cells express tight junction proteins and penetrate gut epithelial monolayers to sample bacteria. *Nat Immunol.* 2001;2(4):361–367. doi:https://doi.org/10.1038/86373

Edelblum KL, Turner JR. The tight junction in inflammatory disease: communication breakdown. *Curr Opin Pharmacol.* 2009;9(6):715–720. doi:https://doi.org/10.1016/j.coph.2009.06.022

https://www.sciencefocus.com/the-human-body/is-swallowing-your-own-phlegm-harmful/ *source for over 1 liter of mucus swallowed per day*

Blander JM. Death in the intestinal epithelium-basic biology and implications for inflammatory bowel disease. *FEBS J.* 2016;283(14):2720–2730. doi:https://doi.org/10.1111/febs.13771 *source for lifespan of intestinal epithelial cells.*

https://www.sciencedaily.com/releases/2016/07/160712142633.htm *source for 10 million epithelial cells per gram of feces*

Chapter 5

Source for relative amount membrane of SER to plasma: Adapted from MBOC, 5th ed. p. 697.Table 12-2.

Schulze RJ, Schott MB, Casey CA, Tuma PL, McNiven MA. The cell biology of the hepatocyte: A membrane trafficking machine. *J Cell Biol.* 2019;218(7):2096–2112. doi:https://doi.org/10.1083/jcb.201903090

Simmons KB, Haddad LB, Nanda K, Curtis KM. Drug interactions between non-rifamycin antibiotics and hormonal contraception: a systematic review. *Am J Obstet Gynecol.* 2018;218(1):88–97.e14. doi:https://doi.org/10.1016/j.ajog.2017.07.003

Agarwal D, Udoji MA, Trescot A. Genetic Testing for Opioid Pain Management: A Primer. *Pain Ther.* 2017;6(1):93–105. doi:https://doi.org/10.1007/s40122-017-0069-2

Shimada T. Inhibition of Carcinogen-Activating Cytochrome P450 Enzymes by Xenobiotic Chemicals in Relation to Antimutagenicity and Anticarcinogenicity. *Toxicol Res.* 2017;33(2):79–96. doi:https://doi.org/10.5487/TR.2017.33.2.079

Alzahrani AM, Rajendran P. The Multifarious Link between Cytochrome P450s and Cancer. *Oxid Med Cell Longev.* 2020;2020:3028387. Published 2020 Jan 3. doi:10.1155/2020/3028387

Ricquier D. Uncoupling protein 1 of brown adipocytes, the only uncoupler: a historical perspective. *Front Endocrinol (Lausanne).* 2011;2:85. Published 2011 Dec 28. doi:10.3389/fendo.2011.00085

Grundlingh J, Dargan PI, El-Zanfaly M, Wood DM. 2,4-dinitrophenol (DNP): a weight loss agent with significant acute toxicity and risk of death. *J Med Toxicol.* 2011;7(3):205–212. doi:https://doi.org/10.1007/s13181-011-0162-6

Alexopoulos SJ, Chen SY, Brandon AE, et al. Mitochondrial uncoupler BAM15 reverses diet-induced obesity and insulin resistance in mice. *Nat Commun.* 2020;11(1):2397. Published 2020 May 14. doi:10.1038/s41467-020-16298-2

Tiniakos DG, Kandilis A, Geller SA. Tityus: a forgotten myth of liver regeneration. *J Hepatol.* 2010;53(2):357–361. doi:https://doi.org/10.1016/j.jhep.2010.02.032

Chapter 6

Alberts B, Johnson A, Lewis J, et al. Molecular Biology of the Cell. 4th edition. New York: Garland Science; 2002. Fibroblasts and Their Transformations: The Connective-Tissue Cell Family. Available from: https://www.ncbi.nlm.nih.gov/books/NBK26889/

Tschumperlin DJ. Fibroblasts and the ground they walk on. *Physiology (Bethesda)*. 2013;28(6):380–390. doi:https://doi.org/10.1152/physiol.00024.2013

Deshmukh SN, Dive AM, Moharil R, Munde P. Enigmatic insight into collagen. *J Oral Maxillofac Pathol*. 2016;20(2):276–283. doi:https://doi.org/10.4103/0973-029X.185932

Nuytinck L, Freund M, Lagae L, Pierard GE, Hermanns-Le T, De Paepe A. Classical Ehlers-Danlos syndrome caused by a mutation in type I collagen. *Am J Hum Genet*. 2000;66(4):1398–1402. doi:https://doi.org/10.1086/302859

Saito M, Marumo K. Effects of Collagen Crosslinking on Bone Material Properties in Health and Disease. *Calcif Tissue Int*. 2015;97(3):242–261. doi:https://doi.org/10.1007/s00223-015-9985-5

Schultz GS, Chin GA, Moldawer L, et al. Principles of Wound Healing. In: Fitridge R, Thompson M, editors. Mechanisms of Vascular Disease: A Reference Book for Vascular Specialists [Internet]. Adelaide (AU): University of Adelaide Press; 2011. 23. Available from: https://www.ncbi.nlm.nih.gov/books/NBK534261/

Wong T, McGrath JA, Navsaria H. The role of fibroblasts in tissue engineering and regeneration. *Br J Dermatol*. 2007;156(6):1149–1155. doi:https://doi.org/10.1111/j.1365-2133.2007.07914.x

Chapter 7

Kanda H, Ling J, Tonomura S, Noguchi K, Matalon S, Gu JG. TREK-1 and TRAAK Are Principal K+ Channels at the Nodes of Ranvier for Rapid Action Potential Conduction on Mammalian Myelinated Afferent Nerves. *Neuron*. 2019;104(5):960–971.e7. doi:https://doi.org/10.1016/j.neuron.2019.08.042

https://teaching.ncl.ac.uk/bms/wiki/index.php/Na%2B/K%2B_ATPase_pump: cites Alberts for the 30% of ATP used for Na+/K+ ATPase)

Daneshjou K, Jafarieh H, Raaeskarami SR. Congenital Insensitivity to Pain and Anhydrosis (CIPA) Syndrome; A Report of 4 Cases. *Iran J Pediatr*. 2012;22(3):412–416.

Chapter 8

Sampath SC, Sampath SC, Millay DP. Myoblast fusion confusion: the resolution begins. *Skelet Muscle*. 2018;8(1):3. Published 2018 Jan 31. doi:10.1186/s13395-017-0149-3

Flann KL, LaStayo PC, McClain DA, Hazel M, Lindstedt SL. Muscle damage and muscle remodeling: no pain, no gain?. *J Exp Biol*. 2011;214(Pt 4):674–679. doi:https://doi.org/10.1242/jeb.050112

Pallafacchina G, Blaauw B, Schiaffino S. Role of satellite cells in muscle growth and maintenance of muscle mass. *Nutr Metab Cardiovasc Dis*. 2013;23 Suppl 1:S12–S18. doi:https://doi.org/10.1016/j.numecd.2012.02.002

Verbrugge SAJ, Schönfelder M, Becker L, Yaghoob Nezhad F, Hrabě de Angelis M, Wackerhage H. Genes Whose Gain or Loss-Of-Function Increases Skeletal Muscle Mass in Mice: A Systematic Literature Review. *Front Physiol*. 2018;9:553. Published 2018 May 22. doi:https://doi.org/10.3389/fphys.2018.00553

Yeung SSY, Reijnierse EM, Pham VK, et al. Sarcopenia and its association with falls and fractures in older adults: A systematic review and meta-analysis. *J Cachexia Sarcopenia Muscle*. 2019;10(3):485–500. doi:https://doi.org/10.1002/jcsm.12411

Amthor H, Macharia R, Navarrete R, et al. Lack of myostatin results in excessive muscle growth but impaired force generation [published correction appears in Proc Natl Acad Sci U S A. 2007 Mar 6;104(10):4240]. *Proc Natl Acad Sci U S A*. 2007;104(6):1835–1840. doi:https://doi.org/10.1073/pnas.0604893104

Rybalka E, Timpani CA, Debruin DA, Bagaric RM, Campelj DG, Hayes A. The Failed Clinical Story of Myostatin Inhibitors against Duchenne Muscular Dystrophy: Exploring the Biology behind the Battle. *Cells*. 2020;9(12):2657. Published 2020 Dec 10. doi:10.3390/cells9122657

Chapter 9

Fuentes N, Silveyra P. Estrogen receptor signaling mechanisms. *Adv Protein Chem Struct Biol*. 2019;116:135–170. doi:https://doi.org/10.1016/bs.apcsb.2019.01.001

Quirke VM. Tamoxifen from Failed Contraceptive Pill to Best-Selling Breast Cancer Medicine: A Case-Study in Pharmaceutical Innovation. *Front Pharmacol*. 2017;8:620. Published 2017 Sep 12. doi:https://doi.org/10.3389/fphar.2017.00620

Khosla S, Oursler MJ, Monroe DG. Estrogen and the skeleton. *Trends Endocrinol Metab*. 2012;23(11):576–581. doi:https://doi.org/10.1016/j.tem.2012.03.008

Scholes D, Ichikawa L, LaCroix AZ, et al. Oral contraceptive use and bone density in adolescent and young adult women. *Contraception*. 2010;81(1):35–40. doi:https://doi.org/10.1016/j.contraception.2009.07.001

Rochira V, Carani C. Aromatase deficiency in men: a clinical perspective. Nat Rev Endocrinol. 2009 Oct;5(10):559–68. doi: https://doi.org/10.1038/nrendo.2009.176. Epub 2009 Aug 25. PMID: 19707181.

Chapter 10

Teraguchi S, Saputri DS, Llamas-Covarrubias MA, et al. Methods for sequence and structural analysis of B and T cell receptor repertoires. *Comput Struct Biotechnol J.* 2020;18:2000–2011. Published 2020 Jul 17. doi:https://doi.org/10.1016/j.csbj.2020.07.008

Nemazee D. Mechanisms of central tolerance for B cells. *Nat Rev Immunol.* 2017;17(5):281–294. doi:https://doi.org/10.1038/nri.2017.19

van Delft MAM, Huizinga TWJ. An overview of autoantibodies in rheumatoid arthritis. *J Autoimmun.* 2020;110:102392. doi:https://doi.org/10.1016/j.jaut.2019.102392

Lutterotti A, Hayward-Koennecke H, Sospedra M, Martin R. Antigen-Specific Immune Tolerance in Multiple Sclerosis-Promising Approaches and How to Bring Them to Patients. *Front Immunol.* 2021;12:640935. Published 2021 Mar 22. doi:https://doi.org/10.3389/fimmu.2021.640935

Sykes M, Sachs DH. Transplanting organs from pigs to humans. *Sci Immunol.* 2019;4(41):eaau6298. doi:https://doi.org/10.1126/sciimmunol.aau6298

Chapter 11

Yang, M., Rito, T., Metzger, J. *et al.* Depletion of aneuploid cells in human embryos and gastruloids. *Nat Cell Biol* **23,** 314–321 (2021). https://doi.org/10.1038/s41556-021-00660-7

Schmerler S, Wessel GM. Polar bodies--more a lack of understanding than a lack of respect. *Mol Reprod Dev.* 2011;78(1):3–8. doi:https://doi.org/10.1002/mrd.21266

Vissing J. Paternal comeback in mitochondrial DNA inheritance. *Proc Natl Acad Sci U S A.* 2019;116(5):1475–1476. doi:https://doi.org/10.1073/pnas.1821192116

Ogawa H, Ono Y, Shimozawa N, et al. Disruption of imprinting in cloned mouse fetuses from embryonic stem cells. *Reproduction.* 2003;126(4):549–557. doi:https://doi.org/10.1530/rep.0.1260549

Demetriou C, Abu-Amero S, Thomas AC, et al. Paternally expressed, imprinted insulin-like growth factor-2 in chorionic villi correlates significantly with birth weight. *PLoS One.* 2014;9(1):e85454. Published 2014 Jan 15. doi:10.1371/journal.pone.0085454

Butler MG. Genomic imprinting disorders in humans: a mini-review. *J Assist Reprod Genet.* 2009;26(9–10):477–486. doi:https://doi.org/10.1007/s10815-009-9353-3

Chapter 12

Khawar MB, Gao H, Li W. Mechanism of Acrosome Biogenesis in Mammals. *Front Cell Dev Biol*. 2019;7:195. Published 2019 Sep 18. doi:https://doi.org/10.3389/fcell.2019.00195

Rahman MS, Kwon WS, Pang MG. Calcium influx and male fertility in the context of the sperm proteome: an update. *Biomed Res Int*. 2014;2014:841615. doi:https://doi.org/10.1155/2014/841615

Teves ME, Guidobaldi HA, Uñates DR, et al. Molecular mechanism for human sperm chemotaxis mediated by progesterone. *PLoS One*. 2009;4(12):e8211. Published 2009 Dec 8. doi:https://doi.org/10.1371/journal.pone.0008211

Rengan, A.K., Agarwal, A., van der Linde, M. *et al.* An investigation of excess residual cytoplasm in human spermatozoa and its distinction from the cytoplasmic droplet. *Reprod Biol Endocrinol* **10,** 92 (2012). https://doi.org/10.1186/1477-7827-10-92

Rojansky R, Cha MY, Chan DC. Elimination of paternal mitochondria in mouse embryos occurs through autophagic degradation dependent on PARKIN and MUL1. *Elife*. 2016;5:e17896. Published 2016 Nov 17. doi:https://doi.org/10.7554/eLife.17896

https://www.avclub.com/every-one-is-sacred-17-songs-about-sperm-1798239104

McGoldrick LL, Chung JJ. Stopping sperm in their tracks. *Elife*. 2020;9:e55396. Published 2020 Feb 27. doi:https://doi.org/10.7554/eLife.55396

Chapter 13

Kobayashi S, D, Malachowa N, DeLeo F, R: Neutrophils and Bacterial Immune Evasion. J Innate Immun 2018;10:432–441. doi: https://doi.org/10.1159/000487756

Wang J. Neutrophils in tissue injury and repair. *Cell Tissue Res*. 2018;371(3):531–539. doi:https://doi.org/10.1007/s00441-017-2785-7

Sender R, Fuchs S, Milo R. Revised Estimates for the Number of Human and Bacteria Cells in the Body. *PLoS Biol*. 2016;14(8):e1002533. Published 2016 Aug 19. doi:10.1371/journal.pbio.1002533

Coombes, J., Powrie, F. Dendritic cells in intestinal immune regulation. *Nat Rev Immunol* **8,** 435–446 (2008). https://doi.org/10.1038/nri2335

Chapter 14

Hanahan D, Weinberg RA. Hallmarks of cancer: the next generation. *Cell*. 2011;144(5):646–674. doi:https://doi.org/10.1016/j.cell.2011.02.013

Tomasetti C, Li L, Vogelstein B. Stem cell divisions, somatic mutations, cancer etiology, and cancer prevention. *Science*. 2017;355(6331):1330–1334. doi:https://doi.org/10.1126/science.aaf9011

Corso G, Figueiredo J, De Angelis SP, et al. E-cadherin deregulation in breast cancer. *J Cell Mol Med*. 2020;24(11):5930–5936. doi:https://doi.org/10.1111/jcmm.15140

Xie Y, Li P, Gao Y, et al. Reduced E-cadherin expression is correlated with poor prognosis in patients with bladder cancer: a systematic review and meta-analysis [published correction appears in Oncotarget. 2019 Apr 5;10(27):2657]. *Oncotarget*. 2017;8(37):62489–62499. Published 2017 Aug 4. doi:https://doi.org/10.18632/oncotarget.19934

Wang P, Zhu Z. Prognostic and Clinicopathological Significance of E-Cadherin in Pancreatic Cancer Patients: A Meta-Analysis. *Front Oncol*. 2021;11:627116. Published 2021 Apr 12. doi:https://doi.org/10.3389/fonc.2021.627116

Sahai E, Astsaturov I, Cukierman E, et al. A framework for advancing our understanding of cancer-associated fibroblasts. *Nat Rev Cancer*. 2020;20(3):174–186. doi:https://doi.org/10.1038/s41568-019-0238-1

Shi X, Luo J, Weigel KJ, et al. Cancer-Associated Fibroblasts Facilitate Squamous Cell Carcinoma Lung Metastasis in Mice by Providing TGFβ-Mediated Cancer Stem Cell Niche. *Front Cell Dev Biol*. 2021;9:668164. Published 2021 Aug 30. doi:https://doi.org/10.3389/fcell.2021.668164

Condeelis J, Pollard JW. Macrophages: obligate partners for tumor cell migration, invasion, and metastasis. *Cell*. 2006;124(2):263–266. doi:https://doi.org/10.1016/j.cell.2006.01.007

Qian B, Deng Y, Im JH, et al. A distinct macrophage population mediates metastatic breast cancer cell extravasation, establishment and growth. *PLoS One*. 2009;4(8):e6562. Published 2009 Aug 10. doi:10.1371/journal.pone.0006562

O'Donnell JS, Teng MWL, Smyth MJ. Cancer immunoediting and resistance to T cell-based immunotherapy. *Nat Rev Clin Oncol*. 2019;16(3):151–167. doi:https://doi.org/10.1038/s41571-018-0142-8

Ralli M, Botticelli A, Visconti IC, et al. Immunotherapy in the Treatment of Metastatic Melanoma: Current Knowledge and Future Directions. *J Immunol Res*. 2020;2020:9235638. Published 2020 Jun 28. doi:10.1155/2020/9235638

Chapter 15

https://www.scientificamerican.com/article/our-bodies-replace-billions-of-cells-every-day/

Index

A

Acrosome, 67, 68, 70, 71
Action potential, 38, 40, 42
Adherens junction, 80
Adipocyte, 29, 30, 53
Aerobic respiration, 17, 26, 28, 29, 47, 50, 64
Alleles, 63
Anabolism, 27
Aneuploidy, 64, 70
Antigen, 76
Apoptosis, 53, 54, 56–59, 64, 75, 79
Apoptosis figure, 57
Autoimmunity, 56, 59, 77, 79
Autophagy, 70, 71
Axon, 37–41
Axoneme, 68, 69, 71

B

Basal, 23

C

Carcinogen, 27
Carcinoma, 81

Catabolism, 27, 29
Cellular senescence, 17
Centrifugation, 55
Centromere, 63
Chromosomes, 16, 17, 61–65, 68
Cilia, 21, 23, 71
Collagen, 33–36
Columnar, 19, 20
Connexin, 50
Crypts, 15, 16
Cytokines, 73
Cytokinesis, 80
Cytoplasm, 2, 25, 26, 28, 33, 43, 45, 46, 50, 52, 53, 64, 67, 73, 76
Cytoskeleton, 11, 12, 20, 31–33, 40, 43, 45

D

Dendritic cell, 22, 75–77
Depolarization, 38–42
Dermis, 31–34, 85
Differentiation, 10, 14, 15, 53–55, 67
Diploid, 63
DNA polymerase, 16, 17
Dynein, 68–70

E

E-cadherin, 80, 81
Endocytosis, 26
Endothelial cell, 73, 74,
        81, 82
Enucleated, 7, 59
Epidermis, 14, 21, 31, 33, 34
Epidermolysis bullosa simplex
        (EBS), 21, 22
Epithelial-to-mesenchymal transition
        (EMT), 81
Erythropoiesis, 7
Exocytosis, 34, 70
Extracellular matrix (ECM), 31, 33,
        34, 36, 81, 82

F

Fermentation, 47

G

Gated ion channel, 39, 40
Glucagon, 28, 50
Glucose, 4, 21, 23, 26–29,
        50–52
Gradient, 23, 29, 40, 46, 52

H

Haploid, 63
Hemoglobin, 4, 9, 12
Hemolysis, 12
Homologous, 63
Hydrophobic, 10, 11, 39, 52

I

Imprinting, 65
Insulin, 4, 27, 28, 49–52, 59
Integrin, 33
Intermediate filament, 20, 21, 47
In vivo, 2

K

Keratin, 20, 21
Kinesins, 41

L

Lumen, 15, 23, 25, 34
Lysosome, 71, 75

M

Meiosis, 61–65, 67
Metabolism, 25–27, 49
Microfilament, 31–34, 37, 43–47, 63,
        68, 75, 80
Microtubule, 23, 40, 41, 61–63, 68, 69
Mitosis, 63–65
Motor protein, 40, 41, 44, 45,
        61–63, 68, 69
Multipotent, 13, 18
Myelinated, 38, 40
Myoblasts, 43, 44, 46
Myofiber, 43, 44
Myofibril, 43, 46, 47
Myostatin, 46

N

Neutrophil, 73–76
Nucleotides, 33

O

Organelle, 7, 28, 34, 37, 41, 64, 67, 71

P

Paracellular transport, 21, 23
Phagocytosis, 74–76
Plasma membrane, 8, 10–12, 25, 26,
        28, 31–35, 37, 39, 41, 43, 45,
        47, 50–53, 56, 58, 63, 64, 68,
        70, 73, 75, 76, 80

Pluripotent, 13, 18
Polar bodies, 64

R

Reactive oxygen species (ROS), 17
Refractory period, 40
Repolarization, 39
Resting state, 38–40, 42
RNA polymerase (RNAP), 14, 33, 65

S

Satellite cells, 46
Signal transduction, 50, 53
Simple epithelial, 19
Smooth endoplasmic reticulum (SER),
    25, 26, 34
Spermatid, 67
Squamous, 19, 20, 73
Stem cell niche, 15

Stratified epithelial, 19
Subunit, 9
Synaptic cleft, 38, 41

T

Telomerase, 17, 79
Telomere, 17
Thick filament, 45
Thin filament, 45
Tight junctions, 21–23, 33, 73,
    80, 81
Transcellular transport, 23, 26

V

Villi, 15

Z

Zygote, 3, 64, 65

Printed in the United States
by Baker & Taylor Publisher Services